We Europeans?
Media, Representations, Identities

The European Science Foundation (ESF) is an independent, non-governmental organisation of national research organisations.

Our strength lies in the membership and in our ability to bring together the different domains of European science in order to meet the scientific challenges of the future. ESF's membership currently includes 77 influential national funding agencies, research-performing agencies and academies from 30 nations as its contributing members.

Since its establishment in 1974, ESF, which has its headquarters in Strasbourg with offices in Brussels and Ostend, has assembled a host of research organisations that span all disciplines of science in Europe, to create a common platform for cross-border cooperation.

We are dedicated to supporting our members in promoting science, scientific research and science policy across Europe. Through its activities and instruments ESF has made major contributions to science in a global context. The ESF covers the following scientific domains:

- Humanities
- Life, Earth and Environmental Sciences
- Medical Sciences
- Physical and Engineering Sciences
- Social Sciences
- Marine Sciences
- Nuclear Physics
- Polar Sciences
- Radio Astronomy Frequencies
- Space Sciences

March 2008

We Europeans?
Media, Representations, Identities

Edited by

William Uricchio

intellect Bristol, UK / Chicago, USA

Cover photo credit:
Olafur Eliasson
The weather project, 2003
The Unilever Series, Turbine Hall, Tate Modern, London
Monofrequency light, projection foil, haze machine, mirror foil, aluminium, scaffolding
Courtesy neugerriemschneider, Berlin; Tanya Bonakdar, New York;
Photo: Karin Becker

First Published in the UK in 2008 by
Intellect Books, The Mill, Parnall Road, Fishponds, Bristol, BS16 3JG, UK

First published in the USA in 2008 by
Intellect Books, The University of Chicago Press, 1427 E. 60th Street, Chicago, IL 60637, USA

Copyright © 2008 Intellect Ltd

All rights reserved. No part of this publication may be reproduced, stored in a retrieval system, or transmitted, in any form or by any means, electronic, mechanical, photocopying, recording, or otherwise, without written permission.

A catalogue record for this book is available from the British Library.

Changing Media, Changing Europe Series Editors: Peter Golding and Ib Bondebjerg
Cover Design: Gabriel Solomons
Copy Editor: Laura Booth
Typesetting: Mac Style, Beverley, E. Yorkshire

ISBN 978-1-84150-207-6

Printed and bound by Gutenberg Press, Malta.

CONTENTS

Foreword

This volume is the product of a major programme under the title Changing Media – Changing Europe supported by the European Science Foundation (ESF). This programme was the first to be sponsored by both the Social Sciences and the Humanities Standing Committees of the ESF, and this unique cross-disciplinary organization reflects the very broad and central concerns which have shaped the Programme's work. As co-chairpersons of the Programme it was our great delight to bring together many of the very best scholars from across the continent, but also across the disciplinary divides which so often fragment our work, to enable stimulating, innovative, and profoundly important debates addressed to understanding some of the most fundamental and critical aspects of contemporary social and cultural life.

The study of the media in Europe forces us to try to understand the major institutions which foster understanding and participation in modern societies. At the same time we have to recognize that these societies themselves are undergoing vital changes, as political associations and alliances, demographic structures, the worlds of work, leisure, domestic life, mobility, education, politics and communications themselves are all undergoing important transformations. Part of that understanding, of course, requires us not to be too readily seduced by the magnitude and brilliance of technological changes into assuming that social changes must comprehensively follow. A study of the changing media in Europe, therefore, is indeed a study of changing Europe. Research on media is closely linked to questions of economic and technological growth and expansion, but also to questions of public policy and the state, and more broadly to social, economic and cultural issues.

To investigate these very large debates the Programme was organised around four key questions. The first deals with the tension between citizenship and consumerism, that is the relation between media, the public sphere and the market; the challenges facing the media, cultural policy and the public service media in Europe. The second area of work focuses on the dichotomy and relation between culture and commerce, and the conflict in media policy caught between cultural aspirations and commercial imperatives. The third question deals with the problems of convergence and fragmentation in relation to the development of media technology on a global and European level. This leads to questions about the concepts of the information society, the network society etc., and to a focus on new media such as the internet and multimedia, and the impact of these new media on society, culture, and our work, education and everyday life. The fourth field of inquiry is concerned with media and cultural identities and the relationship between processes of homogenization and diversity. This explores the role of media in everyday life, questions of gender, ethnicity, lifestyle, social differences, and cultural identities in relation to both media audiences and media content.

In each of the books arising from this exciting Programme we expect readers to learn something new, but above all to be provoked into fresh thinking, understanding and inquiry, about how the media and Europe are both changing in novel, profound, and far reaching ways that bring us to the heart of research and discussion about society and culture in the twenty-first century.

Ib Bondebjerg
Peter Golding

ACKNOWLEDGEMENTS

Collaborative projects of this kind emerge not only because of the efforts of their authors, but because of the participation of many others who create the conditions necessary for such projects to prosper. Thanks are due in the first place to the European Science Foundation and particularly the Standing Committees for the Social Sciences and for the Humanities. This project emerges from an interdisciplinary programme entitled *Changing Media, Changing Europe*, which sought to advance networking among European scholars researching various aspects of the mass media. Initiated by Peter Golding and Ib Bondebjerg, the programme enabled cross-disciplinary, trans-national, and inter-generational collaborations that have blossomed not only into a series of publications, but into robust professional networks and lasting friendships.

I would like to thank the leaders of the project's other teams – Jostein Gripsrud, Peter Ludes, and Els de Bens – for their insights and support. And I would especially like to thank Ib Bondebjerg, whose wisdom and interventions helped to make herding cats not only possible, but pleasurable.

Special thanks are due as well to the many individuals who aided us on our site visits.

Whether informants, who helped us to design our programme and visits, or specialists who shared their insights with us, or media makers who enabled us to listen and see in new ways... these people were invaluable to the project's success. Their efforts permeate and inform the pages that follow.

Heather Owen provided unflagging logistical support, text editing, and humour, enlivening even the most mundane transactions. Melanie Harrison and her team at Intellect Books struck precisely the right balance between professionalism and patience as they coaxed this book into existence.

Most of all, I would like to thank the participants of 'Team Four' – the authors of the essays included in this volume as well as Daniel Dayan, Kirsten Drotner, Sonia Livingstone, Mirca Madianou, Dominique Mehl, Ulrike Meinhof, and Roberta Pearson, and for too short a time, Carmelo Gartaonaindi and Liesbet van Zoonen. Our five years of semi-annual meetings, collaborative explorations, and late night discussions provided, for me at least, an intellectual and emotional substance too often absent from the professional routines of everyday university life.

I am grateful to our team member Karin Becker for taking the cover photo, and to her and Martina Kupiak for arranging permission to use the image of Olafur Eliasson's 2003 Tate Modern installation, The Weather Project.

Finally, an acknowledgement of a very different nature: the essays' heterogeneous citation systems reflect the academic and national traditions of their authors. Although the redactional default tends towards the uniform, this gesture underscores the concern with identity so central to the project.

William Uricchio
Cambridge/Utrecht

We Europeans? Media, Representations, Identities

William Uricchio

The volume in your hand is the result of a unique constellation of factors. It draws on the combined efforts of a group of European scholars bound together by research interests in media. It reflects a series of urban explorations in sites characterized by contested identities, both remembered and lived, from Berlin to Bilbao, Brussels to Budapest, Istanbul to Palermo. And it has as its backdrop a fitful period of time during which project Europe saw a doubling of member nations...and the formal rejection of a unifying constitution. Academic *flâneurs* of sorts, the authors interrogated local specialists, visited sites of media production, participated in rituals of media consumption, and debated among themselves – fiercely at times – the implications for European identity. Grounded in diverse disciplines spanning the humanities and social sciences, deploying an assortment of methodologies, and reflecting different national and generational intellectual profiles, they brought an appropriately broad array of perspectives and analytic strategies to their task.

Working under the rubric of 'homogenization and diversity: media and cultural identities' as part of the larger European Science Foundation-supported *Changing Media, Changing Europe* project, the team explored the relations between media and identity among the many shifting collectivities, both past and present, that constitute Europe. Europe, of course, is a fast-moving target. Whether conceived as a discursive entity or a set of institutional practices, it flits among accreted meanings, embedded memories, and an ever-changing configuration of borders, affiliations and organizations. It stands as a cultural zone – some might even say desire – as complex in its spatial and administrative logics as in the dynamic flows of its histories, its inhabitants and their symbols. Although it enjoys a certain 'taken-for-grantedness' evident as much in everyday parlance as in regulatory practice as in cartography, a closer look reveals underlying ambiguities and exceptionalisms. Tradition coupled with geography, for example, lead some in the United Kingdom to refer to their continental counterparts as 'Europeans', in the process implicitly excluding themselves; while others on 'the Continent' debate the limits of European expansion as it moves to include nations once perceived to be on the cultural and political periphery of the West, such as Turkey. The accretion of different historical 'Europes', together with the complex overlapping of many current 'Europes' (Schengen, the European Union, the European Community, 'fort Europe', the alphabet soup of the EBU, EMU, EEC, and EDC, and even US Defense Secretary Donald Rumsfeld's post-September 11 'old and new' Europe), challenge any easy definitional assertions.

Our attempts to interrogate European media and identity necessarily struggled with this underlying dynamism. How could we even begin to assess media and cultural identity in Europe when European definitional logics were contested as well as in a state of rapid change? Our solution was relatively simple: to seek out sites where these various tensions were writ large; to explore locations where the scarring of historical contestation was still visible; to embrace the tangible fault-lines, fissures, and ruptures that seemed to emblematize the larger European dynamic. Although our questions ranged far beyond the various locales we visited, our work was deeply informed by the specificities of place, and the crystallizations of memory and performances of identity that we encountered there. Found objects, encounters with local informants, media productions, and the lingering traces of the past served as inspiration, sites of interrogation, and provocations to our theories. Our visits and conversations took place over a five-year period, permitting us to deploy and push our ideas in settings defined by radically different dynamics.

The disciplinary and cultural plurality of the authors both amplified and refracted these insights. Among us were film and television scholars and mass communications specialists, but also linguists, policy analysts, sociologists, and cultural anthropologists. In addition to the authors whose work appears in the current volume, our regular conversation partners included Ib Bondebjerg, Daniel Dayan, Kirsten Drotner, Carmelo Gartaonaindia, Sonia Livingstone, Mirca Madianou, Dominique Mehl, Ulrike Meinhof, and Roberta Pearson.[1] Their experiences and voices, although not explicitly acknowledged in the essays, formed an essential part of our discussions and are evident throughout the pages that follow. It is also worth noting that the biographies of our team members served as a constant reminder of the complications of national and European identity. Our group included an Englishman who resides in Turkey, an American who works in the Netherlands, an English woman who grew up in the US, and many hyphenated identities: Moroccan-French, French-Israeli, Swedish-American, and so on. These complexities, compounded by language sets, educational background, personal circumstances (career patterns, partners, and so on), underscored our approach to questions regarding media, various collectivities and identity in Europe.

Thanks to this plurality of perspectives, our collaborative explorations of media and identity across Europe tempered the explanatory capacities of any one analytic paradigm, any one school of thought. It underscored as well the role of national academic traditions and generation in producing nuances of difference and inflections of meaning even within disciplines. While scholarship, particularly in media-related fields, enjoys a high degree of transnational coherence, academic localization is evident in nationally distinct patterns of translation, clusters of literature associated with particular ideas, trends in intellectual influence, and even the meanings of shared words. This variegation mirrored on a methodological level the very complexities of regional identity that confronted us across a Europe 'united in diversity'.

No small part of the diversity that we encountered both in our mode of study and in the practices we investigated can be traced back to language. A plurality of languages must certainly be counted among Europe's cultural treasures, but it also accounts for many barriers to the exchange and interpretation of ideas. The patterning of language competences is also central to the patterning of media, whether facilitating the communication of diasporic populations across Europe, addressing the needs of sub- or transnational minorities, or aligning with and reinforcing the territorial nation state. A number of essays in this collection explore the tensions and possibilities produced by this enactment of language, media and identity.

But as just suggested, language is also the stock-in-trade of scholars, a finely calibrated instrument of analysis and expression in harmony with the cultural spaces that constitute home. And the double displacement that many of our colleagues experienced when speaking in a foreign language about culturally displaced analytic paradigms infused our meetings with a high degree of reflexivity, actively resonating with the very dynamics under investigation. I mention this because I think the reflexivity of our endeavour was singularly appropriate to the dynamism of the cultural practices and European identities that we explored; and because these twin conditions provide a hermeneutic key to the essays here collected. Our project yielded not only the reflections and publications typical of academic work, but helped to establish common discursive ground across linguistic, cultural and disciplinary divides, while building a robust European network of partners and resources.

Setting the Frame

Consider the case of *Europe Day*, declared by the European Union during the 1985 Milan Summit to be 9 May in honour of the Schuman Declaration (1950). French Foreign Minister Robert Schuman proposed a supranational agency that would manage French and German (BRD) coal and steel production as 'the first concrete foundation of a European federation'. This EU decision replaced the earlier *Europe Day* established by the Council of Europe in 1964 that celebrated Europe's founding on 5 May 1949. And in so doing, it repositioned Europe's defining referent from the 5 May celebration of the defence of human rights, parliamentary democracy and the rule of law, to the 9 May celebration of an economic vision. Although only four days apart, there is a world of difference between these two notions of Europe, accounting both for the ongoing contention over which date is more appropriate and a more general indifference to such celebrations. Given the rather greater success of Europe's conversion to the Euro than its adoption of a constitution, perhaps this shift in referent reveals more than its framers intended. In any case, it underscores the definitional contentions and *realpolitische* strains in determining precisely what constitutes Europe.

The conversations and investigations that form the background of this book took place during a particularly volatile period. During the project's five years, ten new nations and nearly 100 million people joined the Union. A post-Warsaw Pact wave of emigration swept westward, while immigrants from Africa, the Middle East and even Asia continued to seek political and economic stability inside Europe's borders. And, following the response of the Bush administration to the events of September 11, 2001, the weight of American cultural products and political pressure seemed more problematic than ever. Not surprisingly, this cultural mix challenged long-held assumptions and revered myths regarding national and ethnic identities, inspiring a predictable political backlash. But at the same time, this state of flux also helped to mobilize a new sense of Europeanness, a concept that seems most sharply defined when positioned in distinction to the 'other,' be it Islam or CNN. The problem with this scenario, however, was that this process of self-definition turned on a logic of contrast, and therefore required the performative fixing of the 'other'. The ensuing notion of 'Europe' that emerged might best be characterized as reactive: a series of compulsive enactments of identity triggered by each new anxiety.

The various cultures, values and reference points called upon in the enactment of identity provoked multiple and competing defensive moves; they sharpened definitional antagonisms; and they served both those on the outside and inside who thought they could play these dynamics for their own interests. This unstable mix, for example, served as fertile ground for US Secretary of Defense Donald Rumsfeld when he described a political rift in Europe in answer to a question by Dutch journalist Charles Groenhuijsen regarding a lack of European support for the US invasion of Iraq. 'Now, you're thinking of Europe as Germany and France. I don't. I think that's old Europe. If you look at the entire NATO Europe today, the centre of gravity is shifting to the east. And there are a lot of new members. And if you just take the list of all the members of NATO and all of those who have been invited in recently – what is it? Twenty-six, something like that? – you're right. Germany has been a problem, and France has been a problem.'[2] But the larger problem – larger even than antagonists such as Rumsfeld – remains Europe's definitional dynamic.

Probing beneath the surface of the present, one can find the densely patterned tracings of vaguely remembered borders, old trade routes, language areas, ideological, economic and religious zones, and former ethnic enclaves. Any and all of these are in principle subject to activation...and therefore, manipulation, particularly in a setting where a contrastive definitional logic reigns, as it does in Europe. The point has not been missed by those seeking to have their way with the European Union, evident from Mr. Rumsfeld's negotiation strategy to American pressure for ever-increasing incursions on privacy for those Europeans flying to or even over the US. In each case, historical divisions are activated for policy ends, complicating the difficult and ongoing work of framing a constitution and crafting a unified sense of Europe. These potentials argue powerfully for scholars of media and history to engage in an archaeology of forgotten memories, suspended identities and dispersed affiliations in order to demarcate and assess the fault lines so open to exploitation.

The potentials of the remembered (or imagined) past both to disrupt and to define underscore the crucial importance of media as brokers and circulators of representation. The media are, after all, repositories of memory, both public and private; and they play an important role in giving us access to selected aspects of the larger world and in constructing the metaphors we live by. We inhabit a moment – exacerbated by the new digitally networked media – where the strategies for media containment and stability of content long deployed by the nation state have largely collapsed. Deregulation of broadcast markets, satellite and cable distribution, the growth of transnational market cohorts, and the migration of content – whether print, image or sound – across multiple platforms, have combined to weaken the state's ability to contain or shape representation. Containment, of course, has been an issue as long as transnational media such as wire services have existed. But the repertoire of well-established techniques for constructing national memory – and its reciprocal, amnesia – has grown inadequate for its task. Even reframing the problem from the Manichean opposition of 'control or collapse' to the far more nuanced ideal of the public sphere, celebrated for its potentials with regard to informed debate over the construction of representation and memory, only underscores the importance of media as a source and site for informed discourse.

The authors in this collection work from a broad definition of media, including public forms such as film and television broadcasts, newspapers, novels, currency, and monuments, and more idiosyncratic forms such as personal photographs, postcards and even branded mouse pads. Each has distinct logics of production, distribution and reception. Each can be approached as traces, fragments, or details, helping us to understand the persistence of memory, the engagement of emotion, or the contestation of meaning specific to a particular situation. And each can be assessed by different modes of analysis and interpretation. But despite these wide-ranging notions of media and a plurality of methodological entry points, the essays all share an embrace of 'situatedness' in their assessments of media-in-action. Whether exploring the role of media as an element in constructing potentially cohesive identities (the trans-European flow of iconographically rich bits of paper in the form of money; satellite transmissions of Turkish television), or examining the interworkings of history and memory through photographs or monuments, the authors foreground the role of media within particular constellations of time and place.

Palimpsest

Italo Calvino observed that every element of the whole is important; no whole exists without parts, no parts mean anything without reference to a whole. His *Invisible Cities*, with its tantalizing taxonomy – cities and memory, cities and desire, cities and signs, thin cities, trading cities, cities and eyes, cities and names – haunts this project, hovering above it like a guiding spirit.[3] His 'part-whole' assessment seems particularly appropriate for the definitional logics of a Europe spread among cities, regions, languages and memories, both defined by and defining the larger whole. And his readings of imaginary cities echo through the reports of our explorations of real cities. These reverberations will be evident in the essays that follow. But there is a third, and perhaps not so obvious, sense in which Calvino's work informs this collection. Teresa de Laurentis has suggested that *Invisible Cities* is an open work in the sense that it challenges narrative patterning itself, exposing its meaning and its logic, and therefore revealing its power. Calvino replaces the usual organization of narrative elements with his curious taxonomy of cities and their tales, offering an alternative and provocative approach to familiar fictional forms.[4] Although our endeavour is far less radical (collections of essays are, after all, rather familiar in the world of scholarship), the variation among perspectives, methods, and levels of specificity stand at odds with familiar approaches to our topic. They challenge the coherence of any particular master narrative in a way that is generative, rubbing one against another in a manner that can produce meanings outside the intent of any individual author.

In order to gain analytic access to the shifting and sometimes elusive dynamic that characterizes the nexus of media, identities and Europe, the essays in this collection work from two different and complimentary entry points. Some move from textually specific engagements, charting particular media encounters in particular locations at particular times. Others operate from a more abstract vantage point, analyzing and mapping spaces of contestation or consolidation. In the case of textually-specific approaches, the essays explore such issues as media production, representational strategies, and meanings in terms of the collective identities they enable. Whether sited in cities or in transnational flows, the analyses reveal the complex negotiations of history, identity, and everyday media as something of an experiential palimpsest. In the case of those studies that are more concerned with what might be called the spaces of identity, tensions among political, economic, and cultural citizenships are variously set into relief by the vantage points provided by notions of the 'other' (religion, the past, 'America', and so on) and by the sometimes dimly recalled memories underlying and shaping the interplay of media and identity.

Although the relationship between media and identity has been taken up in a number of other scholarly works, these studies have typically been concerned with identity either in the individualistic, psychological sense, or in the social sense of national (or European) specificity. This collection, by contrast, takes as its focus the complexities and contradictions of mobile and situation-bound identities as manifest in media production and consumption, identities that simultaneously contest and construct the notion of Europe. The collection's concern with the notion of 'new collectivities' speaks to its authors' understanding of identities as multiple, shifting, and overlapping. With this as a starting point, the authors seek in various ways to explore the rich interactions and the dynamic processes of identity formation that characterize post-war Europe generally and have been particularly evident over the past decade as Europe physically expands and administratively consolidates.

The notion of the palimpsest seems apt for this collection given its overall concern with accretions, shadows, ruptures, and flows. The deep divides of history, whether memorialized in Berlin's Jewish Museum or enacted through violence by the extreme faction of the Basque separatist movement, are traversed by the coherent flows of currency, television formats, and American cultural presence. Memories of violence or past glory accrete in the forms of monuments and photographs, in the forms of fact and fiction, which themselves acquire new meanings in an ever-changing present. The 'new collectivities' whose representation in, and deployment of, media is central to our study, ebb, flow, and morph across borders and across time. Add to this, the multi-vocal and multi-disciplinary perspective of the authors, and the complexity of the palimpsest, at once coherent, multi-layered, bearing the traces of its own construction, and only partially legible, looms large as a relevant metaphor.

The Essays

Rob Kroes highlights the current map of collective identifications, showing how it overlays older configurations of remembered borders and regions that helped previous generations to locate themselves politically and culturally. Proposing an archaeology of Europe as remembered space, he shows how the act of making these memories legible can set the stage for meaningful affiliation and collective action for Europeans *as* Europeans. Kroes goes on to argue that America – as imagined and represented in the European space – may have affected Europeans' sense of their own continent as a stage for collective action of similar scale, effectively offering a homology for self-perception. He explores the possible consequences of this new scale of self-perception in terms of an implied cosmopolitanism, a form of citizenship in a larger democratic space that transcends the prevailing nation state frameworks for cultural and political affiliation. As a model of multi-national democracy, this cosmopolitan Europe is compared to the American space in terms of its possible proselytizing power. Kroes asks whether the European model, by urging candidates for membership in the Union to democratize as a pre-condition for membership, is not a better way to democratize the world than the alleged neo-Wilsonianism – or what he terms a Wilsonianism in boots – of the current American administration.

Working from the particularities of the sites the team visited, Roger Odin explores the processes by which the various cities attempted to build (or rebuild) their identities inside the new context created by the construction of Europe. Using a mouse pad given by EITB (a Basque-language television broadcasting company), a monument in Palermo, a series of postcards found in the gift shop of the Jewish Museum in Berlin, Odin draws upon the notion of the detail, less in the sense of *particolare* (a small part, a fragment) than as *dettaglio*, something symptomatic for the questions he asks, an intellectual provocation of sorts. Working within a semiological famework, Odin moves from the selection of revealing objects, to their *interpretation*, and finally to a combined assessment of these different analyses as a way of characterizing the *paradigm* inside of which the towns' identities are inscribed. In the case of Palermo, for example, he reflects on Mario Pecoraino's monument for those who died fighting against the mafia: three plates of rusted metal reminiscent of Richard Serra's sculptures. This monument, located in the 'Square of the Thirteen Victims', is not historically innocent, and links to Valenti's (now evacuated) monument to the victims of the repression by the Bourbons. The fights against the Mafia and the Bourbons are thus conflated, signifying a *renaissance* for Sicily. In the process of his explorations, Odin moves across expressive forms, drawing on Laetitia Battaglia's photographs, texts from Pasolini, Sciascia and Camilleri, and several films. He reflects upon and compares these various texts, using them as guides to indicate what is changing in Palermo today: an attempt to shift from the anarchic, ephemeral, tricky production of meaning characteristic of Sicilian communication, toward the paradigm of *immanency*, of presentness, a paradigm where truth exists without any interpretation.

Odin's deeply informed analyses flow from his deceivingly unpretentious stance as a cultural tourist. Karin Becker seizes upon the construction of tourism, and explores a very different dynamic, namely the subtle transformations that enable the exile to return as tourist to a once familiar place. Working with photographs, memories and motives, Becker interrogates the experiences of a new breed of tourist enabled by the political and economic changes that have reopened many European borders. She examines what it means to return as a tourist to a place once regarded as home after many years of being away, a place reshaped by faded memory and complicated by layers of images and accreted media accounts. Her focus is on one exile/tourist's photographs and films of the place that once was home, comparing images from before the period of exile with those made when returning as a tourist. She teases out the actualizations of identities, explores how these are brokered through private photograph collections, and traces the ways that images and memories confront the reality of return. While highly particular in focus, the essay offers important insights into the complexity and mobility of the cultural identities being created and recontextualized in contemporary Europe.

Philip Schlesinger reflects on contemporary literary representations of the condition of exile in terms that are broadly ethnographic, suggesting that there are powerful connections between the condition of exile and the ethical obligation to document what this means. Concerned with exploring how we constitute memory and construct identity, both as individuals and as members of groups, Schlesinger takes up W. G. Sebald's work, located on the borderlands of anthropology, journalism, the memoir and the travelogue. In Sebald's case, these generic crossings leave the reader unsure about whether the text is fact or fiction, autobiography or invention. Arguing that Sebald's writing is ethnographic in style and methodology, Schlesinger finds that it shares close affinities to contemporary anthropological ethnography in its coupling of 'first-hand observation with interviews and with historical data and analysis of texts and imagery'. It intervenes in spaces that are by definition undocumented and undocumentable, while being nevertheless precise and experientially grounded, and capable of reaching truths that defy the limits of fact-bound representation. In making this argument, Schlesinger offers a way to reconcile the delicate balance between history and memory so central to the construction of identity. And in the process, he suggests a way to recover the potency and poignancy of certain literary forms as well as the structural lacunae of the past.

As these essays show, advertisements, monuments, postcards, family photo albums, and novels can all serve as textual sites for interrogating the interplay of memories, histories and identities. But media industries, too, bear the inadvertent traces of collective identities. Jérôme Bourdon draws on the example of American influence in the domain of television production – and the ensuing debates on 'Americanization' – to make a case for a complex set of interactions.

Elements of structural and cultural convergence between American and European television can certainly be considered 'Americanization', and such a term appropriately captures attempts to use the medium as a conduit for cultural values. But it also masks a multi-layered set of processes that includes economic logics, production protocols, and highly divergent motives for localization. Bourdon charts many European nations' fascination with American-style light entertainment and journalism, evident in discreetly imported formats and practices. And he shows how the changing economic logics of certain national television systems deterred the wholesale import of American programming at the same time that deregulation encouraged it, allowing him to complicate the discourse of 'Americanization.' At the same time, he demonstrates how these perceptions led to the stimulation of new European genres of infotainment, talk shows, and reality programming. Role model, informant, wholesaler and fierce competitor, American television – like other fields before it (advertising, opinion polls, print press) – was willingly embraced by Europeans, but for their own reasons and to their own ends, greatly complicating the notion of 'Americanization.'

Kevin Robins deals with a quite different manifestation of transnational television broadcasting as he considers the significance of new media for cultural diversity in the new Europe. If public service broadcasting was central to the institution of national cultures and communities, it follows that the new broadcasting culture must be central to the imagination of the new Europe that is coming into existence. In order to make this argument, Robins focuses his discussion on a single case study: Turkish migrants in western Europe and their relation to media cultures. Working from the concrete viewing experiences and voices of this population, he works against the grain of the national imagination in order to reflect upon how we might deal with cultural diversity more adequately – and respectfully. The point is to think beyond old stereotypes of 'the Turk', and to see what some Turks might have to teach western Europeans about cultural diversity. Robins argues that despite its struggles with the question of diversity, the European Commission limited its policy framework to the need to create an expanded, pan-European market ('television without frontiers'). But this *imaginaire* remained essentially national, and did not address – because it could not recognize – the actual complexity and diversity of Europe. The experience of Turkish television viewers in western Europe offers compelling insights into a much more robust vision.

The flows of American and Turkish television programming across Europe offer Bourdon and Robins very different ways to explore the role of the 'outsider' (whether American or Turkish) as an agent in the performance of identity. In each case, the flow of product and programming has long-term implications for transnational identity formation. Johan Fornäs also deals with the transnational flow of symbolic culture, but through a very different medium, money, in his comparative study of euro (€) coins and banknotes as symbolic texts and media artefacts. Coins and banknotes not only communicate an abstract exchange value, but also throw other meanings into circulation in daily life. Produced by the international system of state national banks, they circulate condensed images of national identities and sociocultural value hierarchies through their carefully chosen designs. Thus, they are widely spread media communicating conventionalized collective identifications that reach deep into daily life by being used by virtually everyone on a daily basis. Fornäs offers a close reading of these signs of economic and cultural value, in relation to current public discourses of national and post-national identity, and to ideas on money and cultural identity from Simmel, Benjamin, Habermas and others. He makes comparisons among value levels, countries, and pre-euro national currencies in order to discern value hierarchies, regional and political patterns, and historical changes. The public and political processes that gave birth to the euro designs show how European Union (EU) institutions, states, economic market actors, designers and citizens interacted to develop new forms of identification across Europe. These micro media of communication and exchange greedily criss-cross national borders, but Fornäs asks to what extent and in which ways do they also produce germs of truly transnational identities?

In her essay on the tradition of European 'films of voyage', Maria Rovisco offers a way to rethink collective identity formation in relation to the flow of people by focusing on questions of boundary crossing and national definition. She is concerned with the conceptualization of notions of boundary, and goes on to propose the view that European 'films of voyage' offer a critique of universalistic conceptions of the 'other' while avoiding the fallacies of cultural relativism. Rovisco's underlying assumption is that the voyager can come close to the 'other' in the course of his or her subjective experience of movement across a seemingly foreign space. Thus the difference between self and 'other' can be negotiated, even if the symbolic boundaries enabling the distinction between 'us' and 'them' remain relatively stable and unchallenged. The key question is whether a collective identity is always constituted *against* the 'other', as suggested earlier in the discussion about the role of contrast in Europe's definitional logics. Analyzing three European 'films of voyage' (*Five Days, Five Nights* [Portugal, 1996, Fonseca e Costa], *The Suspended Step of the Stork* [Greece, 1991, Angelopoulos] and *The Crazy Foreigner* [Romania, Gatlif, 1997]), she investigates on both narrative and visual levels how they deal with questions of boundaries and cultural diversity in relation to different spaces with distinct historical contexts.

The final two essays focus on the nation and its relation to regional identities – particularly as evident in media policy – as a prism through which to reflect on larger European dynamics. Sabina Mihelj offers a close analysis of the definitional logics within one nation: Yugoslavia at the point of its disintegration. Focusing on the role of broadcast media and the high-circulation periodical press, she charts the changes in collective identifications and their relations to territory and memory – changes apparent in the erasure of certain terms (for example, socialist, Yugoslav), and the transformation in meaning of others (East, West, Europe, Balkans, as well as all labels referring to specific republics, nations and nationalities). Mass media, Mihelj argues, were among the central institutions that supported such developments by constituting, perpetuating, or changing the contents and reality status of the narratives and rituals that support collective identifications and spatial units. One of the often-discussed features of the Yugoslav communication space was the weakness of pan-Yugoslav media, and especially the fact that until 1990, Yugoslavia had no common, federation-wide television channel. Instead, most mass media with high circulation and coverage were under the control of the various republics (and thus, potentially, nations). Mihelj traces the implications for identity of the complex, changing and often conflicting media narratives of particular Yugoslav nations as they competed for the status of 'reality', 'truth' and 'history', in the process offering a valuable synecdoche for that larger cohesive unit, Europe.

Finally, Giuliana Muscio explores Sicily and its relations to both Italy and Europe as a way to understand the various tensions underlying the enactment of regionalization. Using a close study of Emanuele Crialese's *Respiro*, she focuses on film production as a way of charting the birth of a 'Sicilian cinema' and assessing its larger identity politics. Muscio argues that the tensions between globalization and reactive forms of localization often encourage the adoption of circumscribed identities – leading to what she describes as cultural insularity. Although local movements are often identified with conservative or traditionalist tendencies, she argues that these movements might also be understood in terms of innovative forms of expression and modes of production. Sicilian film production offers a robust example of *glocalization*'s innovative potentials, both in working methods and themes. Moreover, Sicily's cultural position, between the two cultural macro-identities of Europe and the Mediterranean, locates the concrete expressions of its regional identity in a productive space, particularly if we define Europe through a contrastive logic. Crialese's *Respiro* achieved notable international visibility, won several prizes in different festivals, and was nominated for the European Film Award in 2002. In fact within the Sicilian group, it is the only film produced by an international combination (French-Italian) and with European contributions (Eurimages). As a European production it enjoyed international distribution, thus attracting more attention than usual, even reaching the American market. Its marketing strategies pose interesting questions about the cultural identity of European cinema rooted in the regional.

Together, these essays interweave a mix of media forms (novels, money, films, television programming, monuments), of sites (the region, nation, 'Europe'), of flows (of media, of populations) and markers of identity (language, history, memory, trauma, desire). Although somewhat entropic in appearance, the whole is bound together by an overarching concern with the tensions and fissures underlying the ongoing project of Europe's definition, and by the process of collective exploration and interrogation facilitated by the Changing Media-Changing Europe initiative.

A Final Word

Our decision to take *We Europeans?* as a title merits a brief explanation. The year 1939 saw the publication of two reflections on identity, one from each side of the Atlantic. The *Atlantic Monthly*'s "We Americans: Who We Are – Where We Come From – What We Believe – Wither We Are Going" and Julian Huxley's *We Europeans: A Survey of Racial Problems*, although quite different initiatives, both shared an interrogation of cultural values and a relationship to the identity question provoked by developments in Nazi Germany.[5] The issue of race loomed large in both publications, and with it, not only issues of tolerance and civility, but far more fundamental questions regarding identities and coherence in an environment where patterns of race and ethnicity did not necessarily align with certain prominent discourses. These two works mirrored their national conditions, one populist in appeal and format and the other imbued with traces of hierarchization; each in its own way challenged largely unspoken assumptions regarding identity; and each did so by targeting an issue that continues to function as a fault line in cultural identity. Both titles, *We Americans* and *We Europeans*, have continued to appear on books with various subtitles throughout the intervening years, and authors such as Daniel Boorstin and Thomas Allen, Richard Hill and Tony Kusher have continued to interrogate the issues that divide and that bind.

This collection of essays is both indebted to, and very much a part of, this tradition. Europe remains in a state of dynamic development; media systems and flows continue to transform; and the collectivities – the variously clustered people – at the heart of our endeavour continue to shape-shift. Our attempts to assess these dynamics, to strategize ways of comprehending their implications, and to offer tentative conclusions, have all been made with the awareness that conditions continue to shift. Nevertheless, we hope that the reader finds the sum of our multi-disciplinary parts to be greater than the whole.

Notes

1. Most of these colleagues contributed to the team's first volume, *Audiences and Publics: When Cultural Engagement Matters for the Public Sphere* (Bristol: Intellect, 2005), edited by Sonia Livingstone.
2. 22 January 2003.
3. Italo Calvino, *Invisible Cities* (New York: Harcourt Brace, 1974) originally published as *Le città invisibili* (Giulio Einaudi Editore, 1972).
4. Teresa De Laurentis, 'Semiotic Models, Invisible Cities,' in Harold Bloom, ed., *Italo Calvino* (Philadelphia: Chelsea House Publishers, 2001), p. 47.
5. Atlantic Monthly, *We Americans: Who We Are – Where We Come From – What We Believe – Wither We Are Going* (Boston: Atlantic Monthly C, 1930); Julian Huxley, A. C. Haddon And A. M. Carr-Saunders, *We Europeans: A Survey of Racial Problems* Harmondsworth: Pelican Books Penguin, 1939.

IMAGINARY AMERICAS IN EUROPE'S PUBLIC SPACE

Rob Kroes

Where does Europe end? It is a question of immeasurably greater complexity than the question of where America ends. America as a national entity may extend from sea to shining sea, yet as we also know it projects an image of itself far beyond its national borders. People anywhere in the world can meaningfully connect themselves to inner constructs of what America represents and means to them. With the European Union explosively expanding, now having to digest the presence in its midst of new member states that until the end of the Cold War found themselves under the sway of the Soviet Union, the United States blithely leapfrogs across all the new political borderlines in Europe. Travellers venturing beyond the new eastern border of the European Union find prominent displays of McDonald's golden arches in the Crimea, with a statue of Lenin in the background. It is the further extension of a visual presence of iconic images of American mass culture that have featured prominently in European countries during the Cold War. But it has always been a surface phenomenon overlaying the crackled face of Europe, showing an intricate pattern of fault lines and cultural borderlands that is in perpetual flux. Old Europes have vanished, remembered only by those who once lived in vibrant communities that are no more. In his beautiful book of photographs, *Diaspora: Homelands in Exile,*[1] Frederic Brenner presents pictures of the survivors of the 45,000 Jews who had made Salonika, in Greece, virtually a Jewish city. In 1943 they were sent to Auschwitz where more than 90 per cent of them were murdered. Until the beginning of the past century Salonika had been a kind of Sephardi republic: cosmopolitan, Europeanized, linked with London, Vienna, Belgrade and Istanbul by the Orient Express. As a Jewish city-state it lasted from the expulsion of the Jews

Lenin facing McDonald's Golden Arches, Crimea, 2001. Monique van Hoogstraten, photographer. *Private collection*

from Spain in 1492 to the Greco-Turkish treaty of 1923, when 100,000 Greeks from Anatolia and Asia Minor were settled there by the government in an attempt to Hellenise it and end the dominant role of the Jewish majority. German fascism finished it off altogether.

In one of Brenner's photographs of the Jewish Diaspora, four men form a remnant of what had once been the largest Jewish community in the whole of the Orient. Three are on the right of the picture, one on the left. Three fists, two tattooed with their concentration camp number, grip a wooden post. The man to the left of the post holds his hand, palm open, against his face, his own tattoo also visible. Brenner has asked a number of intellectuals, academics, poets and novelists to write short essays on a selection of the photographs in his book. No one has much trouble describing the three men on the right. As Sidra Dekoven Ezrahi, a professor of comparative literature at the Hebrew University, Jerusalem, writes: 'the face of defiance, vengeance, the will to power..."Never again!" shout these grim voiceless faces and those fisted hands.' But over the man on the left there is great confusion. Are we looking at the binary opposite of the other three – suffering versus vengeance – 'sorrow, humility and compassion,' as Ezrahi writes? Or can one detect a 'twinkle in the eye and the ever so slight hint of a tender smile,' as the poet Ammiel Alcalay suggests? I couldn't say. In any case, this photograph as well as the others, gathered in a 25-year journey by the photographer, all testify to the tragedy of European history. Many Europes, many vibrant communities, have come to an end, leaving hardly a trace other than in the memory of scattered survivors. This suggests one answer to the question of where Europe ends. Many Europes have ended many times over, due to genocide, population transfers, ethnic cleansing, and internecine war. As in the movement of tectonic plates, historic forms of Europe have submerged, molten into oblivion.

But Europe ends in different ways as well. Looked at in a certain way, the map of Europe offers a mosaic of borderlands, of invisible lines separating communities from each other. Regional communities, historically rooted, see their cohesion threatened by restless migration movements or more generally by the wider horizons brought by modernization and globalization. Diasporic communities, of Turks, Moroccans, Algerians, live among members of their host societies, on the margin, in cultural interstices, yet with a proto-cosmopolitan sense of the larger European space that they now inhabit, straddling national borders. Certain established Europes, as defined by those sharing a sense of cultural commonality, draw lines to include or exclude neighbours. They are all lines where certain Europes end and rub shoulders with new Europes struggling to emerge.

As I argued before, overlying all this is the idea at least of a larger Europe, offering a framework for meaningful identification to all denizens of the new, emerging Europe. It is a dream more than a reality at the moment. In a stunning collection of photographs on show in the summer of 2004 in the *Kunsthal* in Rotterdam, Dutch photographer Nicole Segers displayed pictures taken along the new eastern border of the European Union, from Finland all the way down to Bulgaria. They mostly are bleak pictures of people left astray by the turmoil of political change since the end of the Cold War. Segers was accompanied by a journalist friend, Irene van der Linde, who interviewed many of the people who found themselves without bearings. Many are now citizens of the European Union. In one conversation, a Bulgarian fisherman says: 'You feel it if you love someone and you feel it if someone loves you.' The fisherman takes a pause, like an actor on stage. Everyone sits in silence. 'In the case of Europe, I feel nothing.' After these words, all the fishermen at the table raise their glass. 'Nazdrave,' they toast. 'Let us drink to this.' 'Welcome to the end of Europe.'[2]

This end of Europe, where people do not have any feelings about the larger political community that now defines their citizenship, is not only to be found at the extreme eastern border of the Union. People all across Europe feel no meaningful affiliation with the 'New Europe,' and are in anguished search for more meaningful frameworks to define their citizenship. In that sense there are many 'liminal Europes,' situations where Europe dims into irrelevance as it reaches its far borders. There are as yet few overarching emblems helping people to conceive of the larger Europe. There are no potent iconic images in the way that America has projected them onto the European canvass. In what follows I propose to contrast these two situations, the crackled pattern of liminal Europes and the presence across Europe of imaginary Americas.

Needed: an archaeology of Europe as remembered space

In my education as a European I remember one formative moment. I had the good fortune as an undergraduate in political science to find a book on the required reading list – Edward Atiyah's *The Arabs*[3] – that shook my established views of the history of western civilization. I had had the privilege to attend an old-style Dutch *gymnasium* and had read some of the classics from antiquity, such as Homer in Greek or Virgil in Latin, in addition to some of the great works in four modern languages, German, French, English, and of course Dutch. It had left me with a mistaken sense of unilineal evolution from Greek and Roman times to modern European civilization. Thus I had an etymological sense of the modern languages at my command, including those of Germanic origin, as resonant with ancient Greek and Latin. I saw Goethe, Shakespeare and Racine as inspired by the masters of antiquity. Words like Renaissance and Enlightenment only confirmed my reading of western civilization as repeatedly reinvigorating itself by returning to its intellectual and artistic origins. Everything in my high-school education had worked to instil in me this sense of history as a transformational process, continuing in one unbroken line, for all its inner hybridity and the admixtures from other sources. My sense and that of many others was the unreflected latter-day version of the old myth of the Westward Course of Empire and Civilization, or as it was known in the days that Europe's common language was still Latin, the *Translatio Studii et Imperii*. Taken up by Bishop Berkeley in the eighteenth century and projected onto the canvas of America, his poem would be eagerly adopted by Americans in the nineteenth century as one of the historical justifications for their westward expansion in what they saw as their manifest destiny.[4] From Virgil to Berkeley there was this sense of an unbroken line of western civilization evolving over time, while it travelled ever farther westward. Nothing had prepared me for the reading of Atiyah's book.

He presents Arab civilization, at the time of its greatest flowering, not as something out there, beyond the self-enclosed sphere of a European world immersed in its own process of civilization, but as critically linked to it, in dialogue, in cultural encounters and clashes, nurturing and further enriching a classical heritage, appropriating it before Europeans claimed it as exclusively theirs. Not only does Atiyah interweave the story of Arab civilization with that of European civilization, offering a larger cosmopolitan perspective, he also explodes current conceptual habits that see Arabs as a homogeneous 'Them' versus an equally homogeneous 'Us'. At the time of its greatest geographical reach the Arab Empire held a population of religious, ethnic, and linguistic variety, yet freely intermingling, and fully partaking of the intellectual and cultural ferment in its urban centres. Atiyah forever changed my mental map of the history of what we now call western civilization, of its locale as much as of its agency. Europe as an organizing idea, conflating a geological landmass with the stage on which western civilization unfolded, would henceforth be a blur rather than offer a clear focus.

Much the same story could be told about the Ottoman Empire, successor to the Arab era of cultural and political predominance. Centred in one of the great European cities, which under the name of Constantinople had for a thousand years been a cultural haven in the history of Christendom, in addition to being a cosmopolitan crossroads, the Ottomans, confusingly and tellingly, invaded it from Europe, entering the city through its West gate, and renamed it

Istanbul. They appropriated its rich cultural landscape rather than razing it to the ground, in marked contrast to the pillage and desecration of a then Orthodox Christian Constantinople at the hands of Frankish crusaders in 1204. Under the Ottomans Christian iconography was plastered over, not iconoclastically smashed to pieces. In the cultural sedimentation of history a new layer was added, like a new coat of paint. In cultural syncretism mosques were built emulating the grand structure of the Aya Sophia. Again, artisans, artists and intellectuals from inside the realm as well as from outside, in fact from all over the larger Mediterranean world, flocked there to contribute to Ottoman civilization. Ironically, with the Ataturk turn towards a radical western secularism and nationalism, seeking to westernize Turkish society following the break-up of the Ottoman Empire, the Aya Sophia was decommissioned as a mosque and restored to its former Christian symbolism. Frescoes and mosaics have been uncovered, and now sit along the later Muslim iconography honouring Mahomet and the first four Khalifs. It is a lasting memento to the history of related, though rival, civilizations washing across each other, in an ongoing ebb and flow.

As happens so often, rivals locked in combat in the end turn out to resemble each other. Within the Christian world the mirror image of the two great Muslim empires is without a doubt the Austro-Hungarian Empire, deriving its cultural sense of itself from its long-time struggle against the expansionist Ottomans as its cultural Other. Yet the resemblance is striking in terms of the multi-ethnic cultural vibrancy, centred in the Austrian case on its seat of empire, Vienna. Nor was the dividing line ever very neat in religious terms. Muslims and Christians lived alongside each other on either side of the line, although tolerance of religious diversity may have been greater under the Ottomans. If tolerance is a virtue claimed on behalf of European civilization, which of the two empires then was more European?

The question is meant to be more than merely flippant. In Christianity's defence against two successful Muslim empires, there was always an exclusionist reading of its cause, a driving sense of religious purity. It fired the fervour of the Crusades; it was behind the Spanish *riconquista*, as much as it inspired later religious wars on European soil. Compared to the religious live-and-let-live attitude in the Muslim empires, what a sorry sight it is to see the successful Christian reconquest of the Iberian peninsula seamlessly blend into the expulsion of the Jews. Nor were the Spaniards alone in this endeavour. In a letter to the French king, the Dutch humanist Erasmus, trying to flatter him, complimented him on making France free of Jews. Elsewhere as well, anti-Semitism served the purpose of creating the necessary Other for cementing cultural homogeneity, around versions of Christianity first, and notions of the nation later. Slowly but surely much of Europe began to harden around lines of social and cultural exclusion. Alien communities were hounded out, their places of worship demolished, their graveyards ploughed under.

Amazingly, after all these years, Europeans still could not believe their eyes when this same logic of ethnic cleansing attended the break-up of yet another multi-ethnic state on European soil, Yugo-Slavia. It was the latest frantic attempt at cleansing the map of Europe from traces reminiscent of earlier forms of communal and cultural life. Photographs and television footage of Sarajevo, showing a multi-ethnic and cosmopolitan place in the grand Austro-Hungarian tradition being shelled by surrounding Serb artillery, left European viewers speechless and powerless. Pictures of concentration camps, showing emaciated inmates clutching the fence that held them captive, evoked instant associations with Nazi atrocities that Europeans might have hoped were forever in the past. The pictures and the associations they called forth, triggered collective memories and provided a ready historical context for the interpretation of what was going on in Bosnia. Yet it took unconscionably long for the West actively to intervene. Impotent anger was the first response. As Barbie Zelizer points out in this context, there is a paradox in this conflation of atrocities remembered and atrocities recently perpetrated. 'The insistence on remembering earlier atrocities may not necessarily promote active responses to new instances of brutality. (I have) argued that the opposite, in fact, may be true: we may remember earlier atrocities so as to forget the contemporary ones.'[5] I am not sure whether I agree with her last point. As I remember it, it was precisely the fact that associations with Nazi atrocities were triggered that gave contemporary witnesses a sense of the enormity of what was going on, and created a large popular pressure in Europe and the United States not idly to stand by but to intercede and stop the brutality. Other more shameful memories came back. After all, nothing had been done during World War II to stop or hinder the Holocaust while it was going on. The photographs that bore ample witness to the atrocities that had taken place were all taken after the fact. Now, in Bosnia, photographs showed atrocities as they were going on. Now was the time to act. In the end, things were brought to a halt before they had had a chance to run their full dismal course.

The Balkan tragedies of the 1990s may have shown demonic dimensions which may make them seem to stand apart as a throwback to a past we mistakenly thought we had forever put behind us, yet the sorting out of European populations along lines of ethnic and cultural purity proceeds apace across the map of Europe, with greater or lesser violence. It ranges from the Basque country, Ulster, Corsica, and Brittany all the way to the successor states of the former Soviet empire. There is much unfinished business on the agenda of cultural purity and homogeneity.

The logic inherent in all this may be of European vintage, centring for the last two centuries on the purity of the nation or of sub-national entities like cultural regions. But it has proved contagious. Thus, in the course of a mere one hundred years following the break-up of the Ottoman Empire, Turkey, as one of its successor nation states, lost much of its cosmopolitan diversity through the forced expulsion of Greeks, whose settlement in fact predated the advent of the Turks on the peninsula, or through the voluntary emigration of people that had given Istanbul its cosmopolitan flavour. Internal migration from the countryside in Anatolia to the industrial labour market of Istanbul exploded its population while making it, demographically and culturally speaking, ever more Asian (its population at the turn of the twenty-first century was about 25–30 per cent Kurdish). As a result, Turkey's most European city has seen its westernized Turkish bourgeoisie and its international community become an ever smaller minority.

Given this re-ordering of the map of Europe over the last centuries, wilfully erasing remnants and markers of earlier social arrangements, it is small wonder that most people living in Europe see the current map as the natural one. To them, this is how things are and how they have always been. As I mentioned before, as a student, I rid myself of this habit of mind by reading a book about Arab civilization. The book, for me, served the purpose of an archaeological exercise in recovering older memories of European space, suppressed and wilfully forgotten, yet of great value for current debates concerning what Europe is and isn't. High-school textbooks across Europe need to be rewritten with a view to unsettling the pernicious presentism of people's ideas about Europe. This ideal new textbook would have to offer a tour across European space, forgetting about borders and current lines on the map, and turning it into one, big memory space, a *lieu de mémoire,* the sort of space as it emerges from W.G. Sebald's journeys across time and space, stumbling upon triggers of memory, bringing back voices long gone silent, shimmering faces emerging from mists before being enveloped by them once again. We need to construct Europe as its own underworld, with ghosts wandering about, demanding to be heard. We need museums, of Jewish history, of colonial history, of regional history, of maps of past Europes, smaller than its landmass or extending far beyond it, museums of population movements, forced or voluntary, of diasporic communities like people of Turkish descent in Germany and the Netherlands finding a new sense of self. We need film and photographs, documenting Europes which have long since vanished. We need interviews with people who now live where others lived before them, in the ruthless succession of populations across the map of Europe, asking them what, if anything, they remember. We need to look at Europe from its liminal points, following the perimeters of contested terrain. And more often than not the liminal points are inside Europe, rather than at its perimeter, as gate-keeping devices for the cultural and civic exclusion of those considered outsiders. We need a map of Europe showing only cultural islands – its pure cultural regions as people see them – and then showing what larger cultural currents wash across these outposts of insulation. We need to turn present-day Europeans into a new audience beholding the pageant of earlier Europes that they should, but often do not, remember.

These would all be necessary exercises in what I called the archaeology of remembered space. Yet we need not only recover the past through uncovering Europe's layered history. Layers are being added right before our eyes, affecting, as they should and for millions actually do, the collective sense of European space, or rather of the many Europes as people now construct them in the wake of the Cold War. In films and writing many creative minds have set out to explore the debris left by the receding tide of Soviet power.

At the end of Anne Applebaum's journey *Between East and West: Across the Borderlands of Europe,*[6] from Kaliningrad on the Baltic through Lithuania, Belarus, Ukraine, Bukovina, Moldova, and Transdniestria to Odessa, she crosses the Black Sea, arriving in the Bosphorus at dawn, and is struck by the colour, energy, and prosperity after months of an ex-socialist drabness of brown and grey: 'Ahead of us gleamed the minarets of Istanbul. I was back in the West.' As Moray McGowan astutely comments,[7] this exemplifies unusually clearly how meaning and identity are relative and constituted through oppositions. Istanbul is usually invoked in western discourse as a bridge between East and West or as a quintessentially Oriental city. But Applebaum, by viewing it as part of a bright, affluent West in contrast to a physically drab and psychologically depressed ex-communist East, locates it, surely with conscious irony, within another, familiar, but very different polarity, that of the Cold War. In Cold War terms, Turkey, NATO's key southeastern flank, belonged to 'Europe' as defined by western powers. Walter Hallstein had, as President of the European Commission at the time of Turkey's Treaty of Association in 1964, declared emphatically that 'Turkey is part of Europe.'[8] This same Hallstein had given his name, as McGowan reminds us,[9] to the West German 'Doctrine' of 1955 which, in a microcosm of western Cold War positions, sought to isolate the German Democratic Republic (GDR) by threatening to sever relations with any third state which recognized it. In certain circumstances, then, Turkey was more western, more 'European,' than East Germany.

Switching to film, there is a masterpiece by the young Swedish director Lukas Moodysson, entitled *Lilya 4-ever* (2002), a bleak and devastatingly powerful study of Lilya, a poverty-stricken teenage girl abandoned in a crumbling town in an unspecified Baltic state. The girl is left behind when her mother leaves, apparently for the United States, with a man she has met through a dating agency. In her wretchedness Lilya finds a friend in a lonely 11-year old boy, Volodya, but then duplicates her mother's betrayal when she meets a smooth-talking young guy who says he can take her away to Sweden: like 'America', another beckoning escape from her pauperized and hopeless life. As it turns out, Sweden offers captivity and slavery rather than liberation. Visually, the director shows us a Sweden that is not all that different from the wasteland of the former Soviet empire. Blocks of high-rise apartments may be in a better state of repair, but otherwise look as grim and forbidding as anything in the East. On arriving in Sweden, Lilya finds that the promised job was a trick. Imprisoned and forced into prostitution, the girl is repeatedly raped, day after day after day, in ways that Moodysson mercilessly shows us from her perspective. Forced prostitution is shown as a horrible reality, one that links the impoverished post-Soviet states and the sex industry in the moneyed West.

If this film conflates 'Europe' and 'America' as paradises of freedom and riches, the connection comes out more strongly in a 1994 film by Italian director Gianni Amelio, *L'America*. It depicts another post-Soviet state, Albania, as a place of unspeakable horror, where anarchy, brutality, and corruption are rampant. The film offers a frightening view of a world devoid of the least trace of civil protection as we value it in the West, a world where every civil institution has collapsed. The final images of escape from this nightmarish landscape by ship present it in the light of a crossing to 'America', although the ship in fact only crosses the Adriatic to Italy. Yet the images are visually reminiscent of ships entering New York harbour, with the steerage passengers out *en masse* to behold the land of promise.

But we need not enter the fringe of the former Soviet empire to get a glimpse of the threats to civil life in Europe. A 1999 film by Spanish director Helena Taberna, *Yoyes*, takes a sobering look at the Leninism and terrorism that characterize the movement for Basque national emancipation. Much as the movement's early appeal was in its use of a language of cultural rights and ethnic self-determination, finding its roots in the larger liberationist enthusiasms of the 1960s, the means it uses in pursuit of its goals are a travesty of the rules of the democratic game. Since democracy came to Spain, ETA has lived a life of denial of its changed environment and has withdrawn into the conspiratorial world-view of its small band of hardened activists. The film tells the true story of a woman member of ETA, a participant in many of its violent actions, who after a period of exile in Paris wants to return to the Basque country and resume a normal civilian life. In retaliation for her defection and denunciation of ETA's totalitarianism, ETA activists execute her in cold blood. It took courage to make the film in the Basque country, given the climate of fear and intimidation that effectively silences the voice of dissent and obscures the plurality of views among the Basque population. If film allows its audience to widen its view, and to look across Europe's borders, a film like *Yoyes* deserves far wider distribution than it actually got. If one of Europe's problems is the creation of truly European audiences, Europe's film and television industries need all the support they can get for the visual representation of Europe's many faces for European audiences.

One further contemporary film by Russian director Alexander Sokurrow, *Russian Ark* (2002), is an example of such a transnational collaborative effort, in terms of its financing and distribution. Ironically, it may have had a greater *succes d'estime* in the United States than in Europe. As almost a cult film, it has run for months in art houses in New York and other big cities in the United States. Although acclaimed by the European press, its public exposure doesn't compare with that in the United States.[10] The film takes us back to the archaeology of Europe as remembered space. Loosely based on Astolphe Custine's classic *La Russie en 1839*, it brings the nineteenth century French nobleman back to life as a ghost, wandering about the premises of the Saint Petersburg Hermitage, but wandering through time as well. At times rooms are filled with present-day Russians, whom the Frenchman engages in conversation, on occasion scolding them for their lack of historical appreciation of paintings that collectively represent European culture. At other moments, in other rooms, we are taken back in time to the days of Catherine the Great, or of the last Czar shown with his family on the eve of the revolution. The final scenes are of a ball in the main ballroom, with the upper crust dancing to music by Glinka, a Russian composer of music in a European vein. With apparent gusto Valery Gergiev, a cosmopolitan star in the world of music, conducts the costumed orchestra. Then, in ominous foreboding of an era coming to its end, we see the hundreds of guests descending the stairs, in a seemingly endless procession. The final shot takes our gaze outside the building, through an open door, into a bleak and nebulous night, as a stark reminder of the dark times ahead, an era in which Russia's infatuation with a bourgeois Europe would be only a distant memory. The film is one astounding, uninterrupted take. The director turns his camera into yet another character in the film, invisible to all, except the French nobleman. They engage in conversation, with the Frenchman going through raptures at the European culture on display, or mimicked in the lavish architecture and interior design of the Hermitage. He tends to disparage Russian culture. It makes his skin itch. Yet in the end he joins the dance, satisfied with the Russian version of the good life in Europe. When all the others leave, he refuses to go along. He would rather continue his ghostly existence as a perennial nostalgic. He takes a little bow to his camera friend and says: 'Goodbye, Europe.'

Custine's *La Russie en 1839* came out at about the same time as Alexis de Tocqueville's *De la démocratie en Amérique*. In the first, the French nobleman had measured Russia by the standards of European high taste and high culture, in the other, he had gone to America to try to fathom what democracy in America, as the wave of a future coming to Europe later, would mean for Europe's cultural landscape. Tocqueville's view was typically ambivalent. He marvelled at the vitality of democratic life in the American Republic. Aware of the old wisdoms concerning the life cycle of republics, from early vigour to decline and eventual demise, he wondered about the secrets of the apparent stability of America's republic. His insights all concerned the forms of social, or 'associative,' life in America rather than the intricacy of its constitutional apparatus. Its political stability was anchored in its social pluralism. Tocqueville was among the first to explore 'civil society,' the social sphere beyond the reach of government control or surveillance, and critically dependent on democratic freedoms. Only they allowed the citizenry to be the agent of its manifold associative activities, ranging from political parties, to churches, schools, and the many other forms of group activities in pursuit of collective interests.

If Tocqueville's grasp of the voluntarism characteristic of democratic life makes him one of the fathers of modern sociology, his exploration of the impact of mass democracy on culture makes him a precursor in a long line of critics of mass culture. Here his views are much more sobering. They held out a warning to Europeans and their more aristocratic views of taste hierarchies and the role of cultural elites. He was keenly aware of the levelling effects of cultural production under democratic conditions, aiming as it did at the lowest common denominator of public taste rather than catering to the tastes of social elites. Culture followed the dictates of majority conformism rather than the intricate games of cultural distinction characteristic of Europe's stratified societies. Both his views of the self-sustaining forces of democracy in America and his views of cultural life under democratic conditions were informed by his sense that what he observed in America was a prefiguration of Europe's future. If Custine's book explored European culture as it was appropriated beyond its eastern frontier, Tocqueville foresaw the advent of mass society as a force undermining European culture from within. As it happened, much of this erosion would be seen by later cultural critics not as indigenous but as a process of Europeans appropriating forms of American mass culture. If Bishop Berkeley had seen a westward course of empire and civilization, the twentieth century, also known as the American century, would show a reverse course. The empire, once it had reached American shores, would now strike back, Americanizing Europe in its turn.

Many have been the discussions in Europe as to what exactly it was that America, seen as the harbinger of Europe's future, held in store. Germany had its *Amerikanismus* debate in the 1920s; France for over two centuries looked on in fascination and trepidation as modernity in its American guise unfolded. Intellectuals in many other countries contributed their views as they observed the American scene.[11] Many travelled there and reported back to their various home audiences; others stayed home trying to fathom the forms of reception and appropriation of the American model, in culture, in economics, and in politics. More often than not the form of these critical exchanges was one of triangulation. Parties engaged in debate on how to structure the future in Europe's various nation states used America as a reference point to define their positions, either rejecting the American model or promoting it for adoption. I have written extensively on these processes of triangulation.[12] Here I propose to take a different approach.

I will look at the presence of iconographic representations of America across Europe, exploring the ways in which it may have affected the sense of European space among Europeans.

American iconography in Europe's public space

I must have been twelve or thirteen, in the early Fifties, when in my hometown of Haarlem in the Netherlands I stood enthralled by a huge picture along the entire rear wall of a garage. As I remember it now, it was my first trance-like transportation into a world that was unlike anything I had known so far. I stood outside on the sidewalk looking in. Not surprisingly, given the fact that this garage sold American cars, the picture on the wall was of a 1950s' American car shown in its full iconographic force as a carrier of dreams rather than as a mere means of transportation. Cars in general, let alone their gigantic American versions, were a distant dream to most Dutch people at the time. Yet what held my gaze was not so much the car as the image of a boy, younger than I was at the time, who came rushing from behind the car, his motion stopped, his contagious joy continuing. He wore sneakers, blue jeans rolled up at the ankles, a T-shirt. His hairdo was different than that of any of my friends, and so was his facial expression. Come to think of it, there must have been a ball. The boy's rush must have been like the exhilarating dash across a football field or a basketball court, surging ahead of others. The very body language, although frozen into a still picture, seemed to speak of a boisterous freedom. Everything about the boy radiated signals from a distant, but enticing world.

This may have been my first confrontation with a wide-screen display of the good life in America, of its energy, its exhilaration, its typical pursuits and satisfactions. Beholding a picture of America in a garage in Haarlem, I was exposed to a representation of life in America in a rare reflection of public imagery that in America had become ubiquitous. Nor was it all that recent there. Even at the depth of the Great Depression, the National Association of Manufacturers (N.A.M.) in typical boosterism had pasted similar images across the nation, advertising 'The American Way' in displays of happy families riding in their cars. Much of the jarring dissonance between these public displays and the miseries of collective life in 1930s' America still applied to Europe in the early 1950s. Those were still lean years. In Haarlem I stood beholding an image that had no visual referent in real life anywhere in Europe. Yet the image may have been equally seductive for Europeans as for Americans. Consumerism may have been a distant dream in post-war Europe, yet it was eagerly anticipated as Europeans were exposed to its American version, through advertising, photojournalism, and Hollywood films.

Now, as images of America's culture of consumption began to fill Europe's public space, they exposed Europeans to views of the good life that Americans themselves were exposed to. To that extent they may have Americanized European dreams and longings. But isn't there also a way we might argue that Europe's exposure to American imagery may have worked to Europeanize Europe at the same time? There are several ways of going about answering this question. It has been said in jest that the only culture that Europeans had in common in the late twentieth century was American culture. Their exposure to forms of American mass culture transcended national borders in ways that no national varieties were ever able to rival. The points of exposure were not necessarily only in public space. Much of the consumption of American mass culture took place in private settings when people watched television in their living rooms, or Hollywood movies in the quasi-private space of the darkened movie theatre. American popular music reached them via the radio or on records and once again made for a formation of audiences assembling in private places, such as homes or dance clubs. This private, or peer-group, consumption of American mass culture does not mean that larger virtual audiences did not emerge across Europe. Far from it. Shared repertoires, shared tastes, and shared cultural memories had formed that would make for quick and easy cultural exchange across national borders among Europe's younger generations. They could more readily compare notes on shared cultural preferences using American examples than varieties of mass culture produced in their own national settings.

Yet this is not what I intend to explore here. There is an area, properly called public space, outside private homes, outside gathering places for cultural consumption that has served across Europe as a site of exposure to American mass culture. Much as it is true that forms of American mass culture, transmitted via the entertainment industry, travel under commercial auspices – are always economic commodities in addition to being cultural goods, to be sold before they are consumed – public space is the area where American mass culture most openly advertised itself, creating the demand if not the desire, for its consumption. In public space, including the press, we find the film posters advertising the latest Hollywood movies, or the dreamlike representations of an America where people smoke certain cigarettes, buy certain cars, cosmetics, clothes. They are literally advertisements, creating economic demand, while conveying imaginary Americas at the same time. They thus contributed to a European repertoire of an invented America, as a realm for reverie, filled with iconic heroes, setting standards of physical beauty, of taste, of proper behaviour. If Europe to a certain extent became 'other-directed,' much like America itself under the impact of its own commercial culture, Europe's significant Other had become America, as commercially constructed through advertising.

If we may conceive of this re-direction of Europe's gaze toward America as a sign of Europe's Americanization, it means an appropriation of American standards and tastes in addition to whatever cultural habits were already in place to direct people's individual quest for identity. Americanization is never a simple zero-sum game where people trade in their European clothes for every pair of blue jeans they acquire. It is more a matter of cultural syncretism, of an interweaving of bits of American culture into European cultural habits, where every borrowing of American cultural ingredients creatively changes their meaning and context. Certainly, Europe's cultural landscape has changed, but never in ways that would lead visiting Americans to mistake Europe for a simple replica of their own culture.

My larger point, though, is to pursue a paradox. As Henry James at one point astutely perceived, it is for Americans rather than Europeans to conceive of Europe as a whole, and to transcend Europe's patterns of cultural particularism. He meant to conceive of it as one cultural canvass of a scale commensurate with that of America as one large continental culture. His aphoristic insight certainly highlights a recurring rationale in the way that Americans have approached Europe, whether they are business people seeing Europe as one large market

Marlboro Country travels – Italian advertisement.

for their products, or post-World War II politicians pursuing a vision of European cooperation transcending Europe's divisive nationalisms. If we may rephrase James's remark as referring to an American inclination to project their mental scale of thought onto the map of Europe, that inclination in its own right may have had a cultural impact in Europe as an eye-opening revision of their mental compass, inspiring a literal re-vision.

Whatever the precise message, the fact that American advertising appeared across European countries exposed travelling Europeans to commercial communication proceeding across national borders, addressing Europeans wherever they lived. More specifically, though, there is a genre of advertising that precisely confronts Europeans with the fantasy image of America as one, open, space. If all American advertising conjures up fantasy versions of life in America, the particular fantasy of America as unbounded space, free of the confining boundaries set by European cultures to dreams of individual freedom, may well have activated the dream of a Europe as wide and open as America. The particular genre of advertising I am thinking of finds its perfect illustration in the myth of Marlboro Country and the Marlboro Man. The idea of tying the image of this particular brand of cigarette to the mythical lure of the American West goes back to the early Sixties and inspired an advertising iconography that has kept its appeal unto the present day (at least in those countries that have not banned cigarette advertising). Over time the photographic representation of the imaginary space of Marlboro Country expanded in size, filling Europe's public space with wide-screen images of western landscapes, lit by a

Dutch poster for Levi's 508.

setting sun, with rock formations glowing in deep red colour, with horses descending to their watering hole, and rugged-faced cowboys lighting up after the day's work had been done. This was a space for fantasy to roam, offering the transient escape into dreams of unbounded freedom, of being one's own free agent. It was hard not to see these images. They were often obtrusively placed, hanging over the crowds in railway stations, or adding gorgeous colour to some of Europe's grey public squares. I remember one prominently placed to the left of the steps leading up to Budapest's great, grey Museum of Art. The show opened right there. One could not miss it.

Advertising across Europe's public space has assumed common forms of address, common routines, and common themes (with many variations). Originating in America, it has now been appropriated by European advertising agencies and may be put in the service of American as well as European products. That in itself is a sign of a transnational integration of Europe's public space. But as I suggested before, the point of many of the stories that advertisements tell refers precisely to space, to openness, to a dreamscape transcending Europe's chequered map. An international commercial culture has laid itself across public space in Europe, using an international language, often literally in snippets of English, and instilling cravings and desires now shared internationally.

Europe's inner contradictions: nationalism versus cosmopolitanism

In current reflections on the ways in which Europe is changing if not evolving, two pairs of buzzwords emphasize the contradictory forces affecting Europe's changes. One pair, cosmopolitanism and transnationalism, focuses our attention on the many ways in which the political affiliations and cultural affinities of Europeans have transcended their conventional frames of reference, away from the nation and the nation state. The other pair, nationalism and localism, stresses the enduring power of precisely such conventional forms of affiliation and self-identification. At the present point in time, with Europe engaged in the Promethean venture of framing a Constitution for the European Union as a result of a dramatic expansion of its scale, a hidebound nationalism and localism is gaining strength. Public opinion in the member states of the Union is increasingly sceptical of the whole project, seeing it as a cultural and economic threat rather than as a promise of a better life for all involved. This may be temporary and transient, a moment's hesitation in the face of a daring leap into a future whose costs may outweigh its benefits. The current economic malaise in much of Europe may in fact lead many ruefully to look back at the days of national sovereignty and the sense of collective control of the national destiny that is now a nostalgic memory. There is a feeling of loss of direction, which in many member states takes people to a renewed reflection on national identity and national culture. Even in countries like the Netherlands where Dutchness has most of the time been more of a 'given' – to use Daniel Boorstin's way to describe the consensual nature of America's political culture[13] – and therefore hardly ever openly contested or argued, has recently become the topic of lively intellectual discussion. The triggers are as much domestic, to do with the increased multicultural nature of Dutch society, as they are European. Yet in the eyes of many the two are interrelated; the increased porousness of national borders is seen as due to the super-imposition of a 'Europe without borders.'

This hidebound view of what is wrong with Europe stands in opposition to views of European developments in the light of cosmopolitanism and transnationalism. German sociologist Ulrich Beck is among those who see transnationalism as the outcome of long-term processes ushering in a stage of Second Modernity; they are processes that have worked to erode the logic of the historical stage of First Modernity, centred on the bonding and bounding force of nationalism in the historical formation of the nation state.[14] Nationalism as a historical project aimed at moulding nations conceived in terms of cultural and political homogeneity, speaking one national language, sharing one cultural identity. Its logic was inherently binary. At the same time as defining insiders, it defined outsiders. These could be strangers in the midst of the 'imagined community' of the nation, subject to a range of forms of exclusion, or they could be literally outsiders, members of other nations, and therefore cultural 'others.' In our age of globalization this binary logic has been relentlessly eroded. Exposed to a worldwide flow of cultural expression, people everywhere have appropriated cultural codes alien to their homogenized national cultures. They have developed multiple identities, allowing them to move across a range of cultural affinities and affiliations. The communications revolution, most recently in the form of the World Wide Web, has made for a freedom of movement between a multitude of self-styled communities of taste and opinion, transcending national borders. A person's national identity is now only one among many options for meaningful affiliation with fellow human beings, triggered at some moments while remaining dormant, or latent, at others.

Political campaign poster, Salzburg, 1997.

One's local roots are now only one of the many signifiers of a person's sense of self. Beck calls this rooted cosmopolitanism. There is no cosmopolitanism without localism.[15]

As Beck also points out, much of this new cosmopolitanism is relatively unreflected; 'banal.'[16] Teenagers affiliating with a transnational youth culture, sharing cultural appetites with untold others dispersed across the globe, are simply consumers of mass culture, unaware of the existential joy of their transnational venture. Banal nationalism is being constantly eroded by the torrent of banal cosmopolitanism in the forms of mass culture that wash across the globe. It is banal *because* it is unreflected, never leading the new cosmopolitans to pause and ponder what happened to their sense of self. Yet, unaware as they may be of the intricate pattern of cultural vectors that guide their cultural consumption, collectively they have worked to cosmopolitanize the nation state from within. Countries like France, Germany, Britain or the Netherlands are no longer nation states but transnational states. Mass culture of course is only one of the forces of change. International migration, the formation of diasporic communities across the map of Europe, and the attendant rise of multiculturalism have also changed the conventional paradigm of the nation state. There is nothing banal here, in the sense of an unreflected cosmopolitanism taking root. Quite the contrary; the anguished consideration of the changed contours of the citizenry is a clear reflection of the concern, shared by many, about what has happened to the idea of the nation. Yet, as Beck argues in *Dissent*, the only way for the European project to go forward is for Europe to become a transnational state, a more defined and complex variant of what its component nations are already becoming.

Much as I agree with this vision of Europe's future, I am struck by the historical myopia in Beck's argument. As he presents his case, Europe's Second Modernity, its age of transnationalism and cosmopolitanism, evolves from Europe's First Modernity, an age whose central logic was that of the nation state. This seems to deny the long historical experience of cosmopolitanism in Europe, of a view of the civilized life centring on what can only be described as European culture. No banal cosmopolitanism here, but the high-minded version of cultural elites producing and consuming a culture that was truly cosmopolitan, transcending the borders and bounds of the nation state. It was always a rooted cosmopolitanism, with European trends and styles in the arts always being refracted through local appropriations, reflecting local tastes and manners. As Kant defined cosmopolitanism, it was always a way of combining the universal and the particular, *Nation und Weltbürger*, nation and world citizenship. This is the lasting and exhilarating promise of European history, in spite of the atrocities committed on European soil in the name of the homogenized nation, marching in lock step, purging itself of unwanted 'others.' The vision of world citizenship, the transcending idea of humanity, has always had to be defended against the other half of Kant's dialectical pair, against the claims on behalf of the nation. This is truly what the post-World War II project of building a new Europe has been all about, to draw on a long European tradition of high-minded cosmopolitanism, inclusive of cultural variety and cultural Others, and internalized by its citizens as a plurality of individual selves.

This is a daunting project. If it succeeds it may well serve as a model to the world, a rival to the American ideal of transnationalism, of constituting a nation of nations. If they are rival models, they are at the same time of one kind. They are variations on larger ideals inspiring the idea of western civilization and find their roots in truly European formative moments in history, in the Renaissance, the Reformation, and the Enlightenment. Larry Siedentop places the formative moment even earlier in time, coinciding with the rise of a Christian view of the universal equality of mankind vis-à-vis God. As he presents it, the formative moment consisted in universalizing a religious view that in Judaism was still highly particularist, claiming an exceptionalist relation between God and the people of Israel.[17] This shared heritage inspired the first trans-Atlantic readings of what the terrorist attack of September 11 signified. It was seen as an onslaught on the core values of a shared civilization. How ironic, if not tragic, then, that before long the United States and Europe parted ways in finding the proper response to the new threat of international terrorism.

In the vitriolic vituperation that now sets the tone of trans-Atlantic exchanges, Americans discard as the 'Old Europe' those countries that criticize the drift of American foreign policy. Robert Kagan contributed to this rising anti-Europeanism in the United States when he paraphrased the dictum that men are from Mars, women from Venus. As he chose to present the two poles, Americans now are the new Martians, while Europeans are the new Venutians. Never mind the gendering implied in his view that Europeans are collectively engaged in a feminine endeavour when they pursue the new, transnational and cosmopolitan Europe. He has a point, though, when he describes the European quest as Kantian, as an endeavour to create a transnational space where laws and civility rule. Yet the Europeans are so self-immersed that they are forgetful of a larger world that is Hobbesian, and is a threat to them as much as to the United States. The European involvement in the larger world tends to emphasize peacekeeping operations rather than pre-emptive military strikes.[18]

Kagan and others tend to forget that it took the United States about a hundred years to find and test its institutional forms and build a nation of Americans from people migrating there from all over the world. It could only have done so turning its back to the world, in self-chosen isolationism, under the protective umbrella of a Pax Britannica. Europe has had only some forty years to turn its gaze inward. During those years it enjoyed in its turn the protection of an umbrella, provided this time by the Pax Americana. This constellation came to an end along with the Cold War. Yet only then could the European construction fully come into its own, conceiving of the new Europe on the scale of the entire continent. It is a tremendous challenge and Europe needs time to cope with it. If successful, though, and this means the inclusion of Turkey, Europe would offer a model to the world, particularly the world of Islam, of a civil and democratic order far more tempting than the imposition of democracy through pre-emptive military invasion. Those in support of what the United States are pursuing in Iraq, blithely call it neo-Wilsonianism. I beg to differ. If there is a neo-Wilsonian promise, it is held by the new Europe, not the current Bush administration.

Notes

1. Frédéric Brenner, *Diaspora: Homelands in Exile* (London: Bloomsbury, 2003).
2. Irene van der Linde and Nicole Segers, *Het einde van Europa: Ontmoetingen langs de nieuwe oostgrens* [The End of Europe: Encounters along the New Eastern Border] (Rotterdam: Lemniscaat, 2004), p. 388.

3. Edward Atiyah, *The Arabs* (Harmondsworth: Penguin Books, 1955).
4. On the history of this myth, see Jan Willem Schulte Nordholt, *The Myth of the West* (Grand Rapids: Eerdmans, 1995).
5. Barbie Zelizer, *Remembering to Forget: Holocaust Memory Through the Camera's Eye* (Chicago and London: The University of Chicago Press, 1998) p. 227.
6. Anne Applebaum, *Between East and West: Across the Borderlands of Europe* (London, 1995) p. 305.
7. Moray McGowan, ''The Bridge of the Golden Horn': Istanbul, Europe and the 'Fractured Gaze from the West' in Turkish writing in Germany,' *Yearbook of European Studies*, 15 (2000): 53–69.
8. Quoted in Udo Steinbach, 'Die Turkei zwischen Vergangenheit und Gegenwart,' *Informationen zur politischen Bildung*, 223 (2. Quartal 1989), p. 43.
9. McGowan, 54.
10. This is true more generally, it is my impression, for European films shown in the United States. When in the United States, in places like Boston, New York, or even a small university town like Bloomington, I found it easier to keep up with recent European films than in my home town of Amsterdam. Important films reached those places, and ran in many cases for many weeks in the various art houses in the area.
11. For a survey of these European debates I may refer the reader to my *If You've Seen One, You've Seen the Mall: Europeans and American Mass Culture* (Urbana/Chicago: The University of Illinois Press, 1996). For a survey of French views of American modernity, see my chapter on the subject in my *Them and Us: Questions of Citizenship in a Globalizing World* (Urbana/Chicago: The University of Illinois Press, 2000).
12. See, for example, Rob Kroes, 'America and the European Sense of History,' in Kroes, *Them and Us*.
13. Daniel Boorstin, *The Genius of American Politics* (Chicago: The University of Chicago Press, 1953).
14. Ulrich Beck, 'Rooted Cosmopolitanism: Emerging from a Rivalry of Distinctions,' in: Ulrich Beck, Nathan Sznaider, and Rainer Winter, eds., *Global America? The Cultural Consequences of Globalization* (Liverpool: Liverpool University Press, 2003). Also: Ulrich Beck, 'Understanding the Real Europe,' *Dissent* (Summer 2003).
15. Beck, 'Rooted Cosmopolitanism,' p. 17.
16. Beck, 'Rooted Cosmopolitanism,' p. 21.
17. Larry Siedentop, *Democracy in Europe* (London: Penguin Books, 2000) 190, 195, 198.
18. Robert Kagan, *Of Paradise and Power: America and Europe in the New World Order* (New York: Knopf, 2003).

Towns in Search of Identity

Roger Odin

Art, affects and politics

Where to start? With the detail, in the sense both of *particolare,* a fragment of the whole, and *dettaglio,* a microelement that I deem to be symptomatic and thought provoking (on this distinction, see Arasse, 1996, pp. 11 – 12). First: I focus on three towns that we visited in the course of our research programme: Bilbao, Palermo, Berlin; three details compared to Europe as a whole, but also details that offer insights into the larger whole, through the way they are trying to solve their identity problems. Second: at the level of the towns themselves, to reproduce this gesture: select one micro-detail. But how? Curiously, the question never arose. I did not have to choose this or that detail; it chose me, it imposed itself on me by surprising, even scandalizing me; it was always opaque, leaving me to wonder why I was looking at this, or at that. After having analyzed the micro-detail, I let my mind roam elsewhere, in literature, cinema, paintings, photographs, architecture. I do this by a process of association, a chain of thoughts; one element leading to another as if by logic. I want both to understand the identity problem and to pinpoint how the town is changing its identity. Then there's a further risk: how to interpret all this? I have to convert my findings into a sense of the cultural paradigm that governs the new identity. This is the challenge that I have set myself in what follows, where I have deployed an approach akin to the semio-pragmatics of *culturality;* culturality rather than culture, since culturality is not a structure (a set of signs) but a process: it is culture in action.

Bilbao and cultural baroque (November 2000)

On the occasion of our visit to the Basque Television station (*eitb*) in San Sebastien we were given several small gifts. I received, amongst other things, a mouse mat for the computer. We can take it as a form of media.

Bathed in blue, the picture on my mouse pad is of three children. Their faces are turned to the sky where they can see a cloud on which the logo and acronym of *eitb* is given, followed by the name in full: *euskal irrati telbista*. The children seem expectant; their faces are serious, solemn even. Time is as if suspended. Our eye is led along the vertical axis, and two different perspectives emerge. The linear perspective is formed by the triangle of children, and leads into the distance towards the object of the children's collective desire (*eitb*) – they are all facing in this direction. The atmospheric perspective, created by the colour blue and its stain of white cloud, gives the impression of depth; this traps the reader who is drawn in further and further, beyond the deep sky into a symbolic depth where they will experience the significance of the meanings involved. Figuratively speaking we are as if before a religious painting depicting scenes of vision and apparition. My mouse mat is like an icon; it has a 'hierophantic' value; it expresses a mystical experience, shows the celestial transcendence that is 'the origin of all things and of law, the signifier of strength and sovereignty' (Damish, 1972, p. 67); this is none other than *eitb* whose logo, in the shape of a bird, strengthens the religious dimension. The vision itself is language, the Basque language to be precise (just as certain mystics saw words written in the skies). On the earthly plane of this vision, there is a bird shelter, which echoes the logo. The message is clear: the children, the future of the Basque nation, are waiting for the Messiah - *eitb* - to descend from the Heavens to earth. Our reading of the image is perfectly self-contained, since our attention does not stray outside the triangular form (the divine triangle) that links the direction of the children's gaze to *eitb* and back again. In this enclosed shape, we have the plastic signifier of an uncompromising message: there is no salvation, there is nothing at all, beyond us and *eitb*. I find this image terrifying. In it, I see everything that I fear in relation to the question of identity: the use of mystical representations to impose an ethnic vision of community, a vision that is all the more terrifying in this case because of its representation of children.

On the day that we arrived in Bilbao, the Spanish press was full of indignation at the murder by ETA activists of a policeman in Barcelona. The film *Yoyes* (2000), directed by a young woman, Helena Taberna, is about these 'bloody outrages of ETA extremism'. *Yoyes* is the story of a young woman who joins ETA where, rapidly, she acquires high-level responsibilities. The film shows how the Spanish government manipulated the GAL (anti-terrorist liberation group) in order to reach ETA and the conflicts inside ETA itself. Following a deadly but mistaken attack on the wrong targets, Yoyes decides to leave ETA, despite the threats issued by certain of her comrades-in-arms. She goes to the USA, then to Mexico. Twelve years later, she settles in Paris, reunited with her husband and child. One day, she learns that the Spanish government has voted an amnesty. Her husband returns to Spain but she stays in Paris because she is uncertain whether the amnesty will apply to her or not, given the mythical quality that her persona had acquired in Spain. Her husband finally manages to convince her to join him, but certain ETA members see her return as a betrayal and she is liquidated on the day of her home village fête.

Although we are told at the beginning that the film story is based on real life, *Yoyes* is no documentary, no activists' film, nor even a historical movie. Not only is the representation of Basque culture and identity reduced to a number of clichés, and to a handful of words of the Basque language, but nothing at all is said about the real reasons why Yoyes joined ETA, and there is no historical reflection, neither on the Basque 'nation', nor on the question of its independence. *Yoyes* is one person's story, and it plays to the full the possibilities of dramatization: the film tries to make us experience first hand the events that it recounts. In this sense, it is more like a work of fiction. Initially, I was quite irritated by this fictionalizing treatment of such a subject, but after having analyzed my mouse mat, I changed my mind. Now, in the film, I see evidence of a three-fold refusal: of the activist's discourse of certainty, of the documentary's informative discourse, and of the historian's discourse of knowledge. *Yoyes* does have a discourse, but it is filtered through fiction, and clearly was not the one that I was expecting. First, it is a discourse of disengagement: *Yoyes* shows that there are times when, in order to be true to oneself, one has to know when to disengage. It is also an anti-ETA discourse, but we have to be clear about the exact nature of the disagreement: the film presents Yoyes' decision to leave ETA not as turning her back on her convictions, not even as the result of political disagreement with the other ETA members, but as a human problem.

The film describes ETA as an organization dominated by individuals who refuse to reflect on their actions. Yoyes, who is an intellectual (she has a Ph.D. in sociology) is not asked to lead the way in thinking about the Basque question, she is asked to lead the armed struggle, and all we see her do is to learn to shoot a revolver. Even worse, when she begins to ask herself questions, she meets with the chilling disapproval of certain of the ETA's members. This is because for ETA, in the way the film portrays the organization, there are no fundamental questions. As with my mouse mat, ETA's members operate at the level of faith. To ask questions is to ask to be eliminated. The ETA of *Yoyes* is an organism animated by fanatics for whom human life does not count, just like the young, handsome blond man (one of the children of my mouse mat, now grown up?) who, having demonstrated his utter lack of concern for the bungled attacks, decides to kill Yoyes. One detail in particular actually inverts religious symbolism: every time that there is a close-up of a little lighter flame (the flames that pilgrims wave in Saint Peter's square when praying to God), ETA commits a murder. ETA, with its blind faith in its own Truth, is on the side of Death. Yoyes, on the contrary, is resolutely on the side of Life.

In addition, the film shows us that ETA and misogyny are part of the same picture. My mouse mat also portrays a certain misogyny. Not only are there two boys for one girl, but the girl is in the background, behind the boys. Whereas the low camera angle and the lighting accentuates the strong chin of the boy's, the girl's features are softer, and her posture more hesitant. Clearly, she is not about to take the destiny of her country into her own hands. We can allow ourselves to think that this misogyny is not entirely alien to Yoyes' final fate: a woman cannot become an ETA leader and go unpunished.

In one sense, the title of the film says all. *Yoyes* highlights the heroine's name, that signifier which is the one that most portrays the human being, as opposed to the acronym ETA, which signifies a dehumanized collective entity. From this perspective, using the register of human emotion to communicate with the viewer comes across as a particularly fitting strategy. By its very structure – fragmented narrative, blurred chronology (the film propels us between different historical moments), more emotion than rational discourse – *Yoyes* is like a *memoir.* It is possible that this structure works better than History or Discourse in making the viewer experience the complexity of a situation such as that of the Basque country. *Yoyes* is a 'film of stimulation': at the end of *Yoyes*, the death of the heroine brutally shook me out my comfortable position as a spectator, and forced me to ask questions. *Yoyes* is a *cri du cœur*; it tells us that 'enough is enough'. It is no coincidence that this should be a woman's cry; a woman who, like her heroine, does not lack courage: to call for an end to violence is sometimes far more risky than the alternative.

In front of our hotel, an art gallery has opened in the dockyard buildings, most of which have been demolished for renovation: La Brocha. It is showing a retrospective of the works of Rafael Ruiz Balerdi (1934–92), an important figure of Basque Country painting. To my eyes, Balerdi's paintings are the exact opposite of my mouse mat: monochrome vs multiple, bright colours; linear planes vs juxtaposition of shapes; emphasis on the Word vs sensations; dematerialization (in the mouse mat, we are invited to forget texture) vs matter (chalk, watercolour, drawings, oils); closed structure vs open, bursting composition; total control vs spontaneity and openness to accidents; purity vs *mélange*; monosemy (to celebrate *eitb*) vs polysemy (Baledri's pictures are resolutely open to interpretation); expectations of divine intervention vs joyful surrender to the wonders of the world (vegetable, mineral, human); myth of origin vs myth of the cosmos; mysticism vs materialism, and so on.

Thus La Brocha's achievement is two-fold: the æstheticization of an industrial space, and the affirmation that identity (Balerdi's work has an unmistakable identity) can be *mélange*, openness, a welcoming of the world diversity. There is every reason to believe that this twin message is entirely intentional on the part of the Bilbao authorities. La Brocha is part of a long-term project whose aim is to identify Bilbao with contemporary art: *viz* the installation on the sea front of galleries designed by Cesar Pelli, the creation of Norman Foster's new underground transport system, the building of a new airport terminal and of the Uribitarte bridge (just outside la Brocha) designed by Santiago Calatrava ...But it is the Guggenheim Museum which best drives this two-fold operation.

Everything that has been written about this building emphasizes its roles as an operator in the transformation of Bilbao from a 'city of steel' (website), a 'large, unattractive industrial city, strictly for lovers of urban poetry' (*Guide du routard,* 1997), to an artistic centre of world significance. Frank O. Gehry's *œuvre* is a gesture of identity that encompasses not only Bilbao, but the whole of the Basque country. The Guggenheim's identity is plural (it is composed not of one sole building but of the juxtaposition of fragments, building an amorphous, varied whole), polymorphous (you can pick out the shape of a boat, an animal or a vegetable) and relational: it is the sum of its relations with the other Guggenheim museums in the world, with the Bilbao's past (its maritime heritage), with the Nature around (the river but also the sky which is reflected in certain planes of the building), and more generally with the vitality of the World (as with Balerdi's paintings, there is something cosmic in it). We can talk about the mythical dimension of its architecture. But the most significant point is the effect of movement that it suggests overall: it is not so much a monument as a *movement* – to quote an expression coined by Francis Ponge (2002, p. 908) to refer to the Beaubourg Centre.

La Brocha, Balerdi and the Guggenheim seem to me to belong to one and the same paradigm; I would call it: *cultural baroque* (Abdallah-Pretceille and Porcher, 1996, p. 66). Although on the surface it appears similar to the postmodern label so frequently used about the Guggenheim, cultural baroque differs radically in pragmatic terms. Whereas postmodernism signifies the 'end of history', the end of the great narratives (Lyotard), cultural baroque opens up a mythical dimension. Where postmodernism signifies the end of *the social,* the loss of meaning and values (Baudrillard, Jameson), cultural baroque is a socio-cultural act, in respect of which we can talk of 'acting culturally' (*agir culturel,* to paraphrase Habermas). Cultural baroque deploys a rhetoric which wants to change the world and, in this specific case here, enable the Basque Country to escape its identity trap by substituting for the single-roots concept of identity an approach to identity defined by relations, transformations, mix and multi-culturality; in a word, the complete opposite of the notion of identity portrayed in my mouse mat.

Palermo: a strategy of immanence (April 2001)

My meanderings in Palermo brought me face to face with the memorial to the dead in the fight against the Mafia, situated on the Square of the Thirteen Victims. The Denis Tissander and Jean Modot guidebook (*Sicile*, Paris, La Manufacture, 2000) tells us that earlier, on the same site, there was a different monument, a sculpture by Valenti. For me, this substitution is very revealing of the ways in which Palermo, and Sicily more broadly, have been changing.

To erect a memorial in homage to those who have died in the struggle against the Mafia in Palermo is without doubt an unprecedented symbolic gesture. Of course, in Palermo and in Sicily there were already many commemorative plaques but, here, the size of the monument, its location, its shape, all signify the importance of this gesture. It is not solely commemorative, but is an affirmative act. In general, memorials are erected after wars, when the war is over, to mark our recognition of those who fought and lost their lives. To erect a memorial in homage to those who died in the struggle against the Mafia is to indicate that this war is now over, or at least to say that the whole of Palermo is on the side of those who died in the fight. In this gesture, we can see the outcome of the policy implemented by Leoluca Orlando, the former mayor of Palermo (he was subsequently elected to the Council of Sicily) who dedicated his term of office to changing the relations between the inhabitants of Palermo and the Mafia; his initiative became known as the 'springtime for Palermo':

> Before [...] the town was literally split in two: on one side were the silent marshes, inside the town, and, on the other, clans intent on killing each other when they where not murdering celebrities to gain supremacy over the neighbourhood. Today, there is real 'contamination' between the heroes and the whole of the population (Ariel F. Dumont collected all Orlando's quotations; see Intern@tif, Friday 1 August 1997)

The site of the memorial was evidently not chosen lightly. The Square of the Thirteen Victims forms one of Palermo's major crossroads; it is at the junction of Cavour Street which leads to the *Palais de Justice* via the Piazza Verdi where the Opera is (the *Teatro Massimo*) and Francesco Crispi Avenue which at this point becomes the street dei Barilai which leads alongside the docks to the port (la Cala). The square borders onto the remains of the Arab district – there is a sign beside the site of an archaeological dig explaining that they are excavating the ruins of a district of Muslim slaves from the tenth century –, and it is located next to the Church of Saint Giorgio dei Genovesi, dating from the sixteenth century, and the Oratory of Santa Zita, masterpiece of baroque by Giacomo Serpotta. This square, thus, is a compact version of Sicily's history.

My immediate thought was that the thirteen victims were victims of the Mafia, but I was wrong: they were killed on the orders of the Bourbons in April 1860, after the discovery of a plot at the convent of Gancia. The violent repression that followed was one of the principal reasons for the intervention of Garibaldi and his Thousand Volunteers. History tells us that this incident led to Sicily becoming part of the new Italy, under the reign of Victor-Emmanuel. The switching of the memorial to the dead in the fight against the Mafia, and Valenti's monument, invites us to embrace these two monuments in the same paradigm. What the substitution tells us is that, for Sicily, liberation from the Mafia is just as important as liberation from the Bourbons. The new monument signifies nothing less than a new birth for Sicily: a *renaissance*.

Mario Pecoraino, a plastics artist from Palermo, built the memorial. Its shape is particularly important with regard to its symbolic function. It is what we could call a 'structure-sculpture'. There are three enormous sheets of metal standing over ten metres high, reaching skywards, and anchored to the ground in such a way that they converge towards the same point, forming a sort of irregular star shape. They pierce the sky with their black mass. Low down on each sheet, roughly at eye level, a horizontal line has been etched; it has been corroded by rust, it is coloured a brown-red – it is bleeding. On one of these lines, the one facing the town (not the sea), we can make out an inscription in relief: AI CADUTI NELLA LOTTA CONTRO LA MAFIA.

To the best of my knowledge, this structure is the only abstract art monument in the entire city. It is clear that Palermo looks far more to its past, and to its traditions, than toward modern art. Tourist brochures tell us about baroque churches and palaces, show us small, local, pretty markets, ornate carriages, and puppet theatres. In choosing a work of avant-gardist style, Palermo's intention was to change its culture of reference. Orlando had very explicitly included this change in his programme for the city: 'We must [...] fight the Mafia on its home turf by taking its place. In other words [...] by constructing new cultural spaces that offer an alternative to Palermo's inhabitants.' Pasolini, Gada, Camilieri, Sciascia and others have each pointed to the two fundamental rules that underpin Sicilian culture, and the Sicilian-style of communication: never speak plainly, never tell the truth. Clearly, the Mafia is at the heart of these strange practices, with its coded language, its mixture of silences and gestures comprehensible only to the initiated; not to mention *omerta* – the refusal to communicate. Nearly all Sciascia's novels deal with this topic: slander, anonymous letters (*À chacun son dû*), forged translations (*Le*

Conseil d'Egypte), '*a baccagliu* recommendations' – 'a jargon where what is said, and what is not said, flattery and threat, the godfather's handshake, the wink from the rogue, are all in some way examples of *artistic* transformations' (*Les paroisses de Regalpetra*) ...This obscurity is part of a way of life in a place where meaning cannot be direct, where the literal is banished and the cultivation of opacity serves as an instrument of power and defence. Numerous texts spell out the risk involved in even contemplating a discourse of truth: Dibella, the witness, is murdered on his doorstep (*Le Jour de la chouette*); Laurana, who took it upon himself to investigate a murder solely out of 'a purely intellectual curiosity' (the idea that solving the mystery would ensure the triumph of Justice, never entered his head) was also murdered (*À chacun son dû*). Sicily is also described as the country of fanciful interpretation, the land of the generalized misreading. In *Du côté des infidèles,* a Sicilian bishop falls victim to the misinterpretation of his behaviour by the authorities that see him as a political threat.

This 'communication *à la sicilienne*' is certainly nothing new. We can detect it in Sicilian baroque where the proliferation of signifiers and symbols convey multiple meanings, where extreme shows of pathos lead to a scrambling of meaning by the sheer excess of sensations. One of the most spectacular examples of this is the 'Villa Palagonia' built by Ferdinando Gravina (1705), a theatre of illusion, a *trompe l'œil,* unveiling an improbable collection of monsters blurring the classification of the species. But the most curious symbol of this particular conception of communication is the island's coat of arms which hails from Antiquity: a Gordon's head in the centre of a circle from which three legs are suspended, with winged heels, in spiral fashion.

Letizia Battaglia's photos are conceived to counter this 'communication *à la sicilienne*'. Since 1990, Battaglia has worked in Palermo alongside Orlando as a town councillor and has launched herself fully into the fight against the Mafia. Her photos (see Battaglia: 1999) constitute a visual account of twenty years of the Mafia, and some of them have been used by the courts to prove collusion between politicians (among whom, Andreotti) and the Mafia. But above and beyond their value as historical documents and legal evidence, Battaglia's photographs are noteworthy for the way in which they invite us to view reality. Shot in black and white, they show plainly, without pretence, the violence and death that, until very recently, were daily diet of the inhabitants of Palermo. Let us look at two photographs taken from her series on Mafia crimes.

In the first photograph, we see a deserted alley between two high walls, where a corpse is lying face down. At the end of the street is a small crowd of curious onlookers. The choice of a wide-angle shot has exaggerated the convergence of the lines drawn by the walls with the effect of drawing our eye to the body that lies in the centre of the alley, in the foreground. Our gaze is then taken in the opposite direction, down from the head (this is a portrait shot) that lies in a pool of blood to the bulky body wrapped up in a pin-stripped coat, ending at the lower leg, bared at the ankle; this constitutes the only white in a predominantly grey image: the flesh of the corpse. In the second photograph, there is a car whose front window on the driver's side has been shattered. Inside sprawls a man, his head jerked back. The effect produced by the car window (a frame within a frame) creates a scopic trap. The contrast between the lit corpse and the darkness of the night reinforces this effect, and it is hard not to be drawn into observing the man in closer detail, since he is as if offered up to us. The signs of death are written on his face: eyes rolled back into his head, and his mouth wide open, a string of blood from his mouth to his chin and down onto his carefully buttoned raincoat. In the foreground we can just make out the waving hands and crackle of the camera flash. On the right is a policeman, moustached, upright, impervious, observing the scene.

Beyond the subject matter and the composition of these photographs which exploit to the full the technique of *monstration* everything – is done to force us to see the corpse – the great strength of these pictures is that they show the result of actions that have already occurred. They invite us to reconstitute the story, but we do so beginning at the end, and this end is the corpse. From this perspective, these photographs function in the exact opposite manner to the films on the Mafia that we are accustomed to, where we are interested in the events and stories leading up to the murders, and not in the bodies themselves: they disappear in the telling. Battaglia's photographs, on the contrary, fix the aftermath of the action: what is most irredeemable. There is nothing to distract us from the image of the dead. Not death, but the dead, since these photos do not show us death in general or abstract terms but specific individuals (often named); people like us, but dead. This seems a good point to emphasize the pertinence of the choice of the photographic medium. To choose the photograph as medium is, faced with 'communication *à la sicilienne'*, firmly to refuse verbiage: to count on the 'savage silence of the photograph' (Mons, 2002, p. 39). Second, it is to counter the frenzy of images: the photograph works by virtue of the unity of the image. Finally, the violence inherent in the act of giving us fixed images of the dead serves to emphasize the violence of the act of killing: 'by petrifying time and space' we have the 'essence of a halt'; by fixing time, the photograph 'forcefully fills our entire view' (Mons, 2002, pp. 30–31). For Roland Barthes (1980, p. 144), 'the photograph can be taken to signify the intrusion in our modern society of an asymbolic Death'. Alain Mons goes further and speaks of the 'madness of a plain death (no ritual, no religion), of which photography is the hallucinatory, visual symptom' (ibid., p. 30). In the eminently religious society that is Sicily, Battaglia's photographs are redolent in their efficiency, the matt quality of their violence.

Battaglia's work aims to promote a mode of communication based on a brutal revealing of the truth: the scandal of the dead at the hands of the Mafia, hoping that this will provoke Palermo's inhabitants into feeling concerned. In an even more direct manner, Mario Pecoraino's memorial constitutes a shock with regard to the 'rules' of 'communication *à la sicilienne*'. As abstract art it runs counter to those practices that favour figuration, narrative and fabulation; as a minimalist piece it turns on its head the inherent over-abundance of baroque. In place of artifice, it insists on the respect of raw matter. In the face of the symbolic and its corollary, over-interpretation, it offers an explicit message that needs no interpretation. In sum, it is a gesture of truth, and the message is the same as that of Battalia's photographs: a statement of the scandal of all these dead, a purely secular statement.

As such, Letizia Battalia's photographs and Mario Pecoraino's memorial take place in what we can call the *paradigm of immanence*. We refer here to all those works whose power lies in the raw statement, in their presence alone. Mario Pecoraino's structure-sculpture echoes thus a whole series of works found the world over (especially Richard Serra's structures based on large upright slabs). Mario Pecoraino's sculpture thus forms part of an international network of pieces that function in their way as signs of truth, spreading from one space to another, transcending distance and countries.

Thus, Palermo makes a statement of its intention both to change its cultural paradigm and to integrate into the world by refusing insularity and the temptations of isolationism. Palermo's communication strategy is similar to that of Bilbao's: using art as a means of opening up to the world, but it is also the opposite. Unlike Bilbao, it is not a question of turning Palermo into a city of art; Palermo is already a city of art. Nor is it a question of using art to engender a proliferation of communication; this is precisely what must be avoided (here, baroque is the enemy). It is more a question of recentring Palermo and Sicily around a limited number of values – the value of truth, of human life – by denouncing the scandal that constitutes the very opposite of these values: intentional death. A number of the decisions taken by the Orlando administration show how this particular value (of human life) is very much at the centre of the work undertaken by Palermo to build an alternative identity for itself. One of the most spectacular of these decisions was to bury in Palermo the American Joseph O'Dell, executed on death row, and to nominate as honorary citizens those on death row throughout the world: the American journalist Mumia Abu-Jamal, the Japanese Kimaru Shugi, the Guyanan Noel Tomas, the Zaïrian Nessy Ngungasa, the Jordanian Amira Salem. Orlando spoke of the 'symbolic aspect of this gesture':

> For many long years, Palermo was seen as the city of the Mafia, that is to say, of death. By protesting against the execution of O'Dell and the hundreds of people awaiting death, we are making a highly symbolic gesture. [...] La Pieuvre's only culture is that of death; we are offering Palermo the culture of life. (Dumont)

By taking this political decision, Sicily is claiming its right to belong to a collectivity defined by its ethics, a collectivity that transcends nations and even continents but which, at the same time, provides for the building of a strong sense of identity, since it is based on clear and stated priorities.

Berlin, or intimate exposure (November 2002)

Following my visit to Berlin, I found two postcards that I had brought back with me from a visit dating from before the fall of the wall (1989). Both feature optical effects – whereby by tilting the card or squinting slightly, the eye moves between two different images on the same card.

The first image is a colour photo of an official building surrounded by green lawns, where Berliners evidently like to promenade. The second image, on the same card, shows the same building, but this time in black and white: the photograph was taken at night and the building is burning; in the foreground we see a crowd from behind; they are watching the flames lighting up the sky and the dome. We are invited to see what the crowd sees. On the reverse of the card is the following caption: *Berlin. The Reichstag. Built by Wallot, 1883. Burned by fascists. Never forget* (my translation). The link that this caption invites us to make between the two images is a classical one of oppositions: the colour photo relates to the happy present, the black and white to the dramatic past. Considered thus, the postcard takes a stance within one of the big debates currently agitating Germany: what to do with the Nazi past? Should we remember? Revisit it? Or maintain our silence; push it away in order to advance into the future? The postcard takes a clear position: never forget. However, we should note that it is the caption that makes me read the card in this way. If we limit ourselves to these two images alone, other readings become possible, for example, the following. I was struck by the similarity between the black and white photo and a photograph of the destroyed Reichstag reproduced in Brian Ladd's *Ghosts of Berlin* (1997, p. 90). But the photo in Ladd's book is from 1945 and the destruction in question is that inflicted by the Soviets. Thus without the caption, we could read the postcard's double image as an anti-Soviet, and not an anti-Nazi discourse.

Turning to the second postcard, the first image, in colour, shows a night scene of a Nazi procession in front of the Brandenburg gate. The twin image, in black and white, shows a desolate landscape devoid of human presence: a row of ruined apartment blocks, doubtless following bombings. The caption explains: *Berlin, 1933–1945. When the multicoloured flags fly, who can foresee when it will end? Fascism never again!* 'Anchored' by its caption, the aim of the image-montage is clear: Nazism led to mass destruction. However, in this case the colour/black and white contrast is not without difficulty. Whereas the black and white image clearly connotes a tragic past in relation to the time when the postcards were produced (1983), the use of colour for the Nazi period raises a number of questions. Clearly, we can imagine that those who designed this card wanted to produce a dramatic effect; nevertheless, such a reading implies a set of contrasts that is not in keeping with the first postcard. If we apply the contrasts that we applied to our reading of the first postcard, the destroyed buildings in black and white are relegated to a dramatic past, and the colour image of the Hitler march to a happy present. In the context of this reading, the dramatic past could thus be that of the destruction of Berlin by the Soviets in 1945, and the image of the Nazi procession of that against which Nazism fought: communism.

The two postcards can thus be read as Germany taking itself in hand again after the Soviet bombings, which would imply that Nazism was integrated into a positive reading of history. We know that this discourse does indeed exist in Germany. It is exemplified in particular by the historian Nolte: 'The only reason why Hitler and the Nazis committed an "Asian" crime was surely because they deemed themselves and their own as potential or real victims of an "Asian" crime. Did the Gulag not outdo Auschwitz? The "murder for class reasons" perpetrated by the Bolsheviks is it not the logical and factual predecessor of the "murder for reasons of race" perpetrated by the Nazis?' (Robin, 2001, p. 79). Of course, the caption accompanying the card makes it perfectly clear that this is not what the images intend to convey, but do we always read the caption, especially when it is printed on the reverse of the card and written in German?

Even if we can argue against the deviant reading on the grounds that it is too random, one thing is certain, which is that the cards shake meaning in the most concrete manner possible: it is difficult, when I hold the cards, to focus on one of the images; I'm always somewhere between the two images, moving with a blink, a slight movement of my head, from the black and white to the colour, from the past to the present, from the present to the past, from the Nazi period to today, and back again. This reversibility of time creates a tension, a tension connected to uncertainties surrounding the management of the problem of the past – what the historians call the *Vergangenheitsbewältigung* (control of the past). There is more. These cards operate as a little portable interactive art installation: it is me, the reader, who produces this temporal coming and going. It is myself who, by handling the cards, alters the meaning of the image-montage, and thereby the meaning of history that they relate. I am involved. Considered in this way, these postcards do not simply show Germany's present hesitancy in relation to its Nazi past; they also display the desire to involve the whole world in its problems (the cards can be sent anywhere in the world, they are mass media par excellence). My story of postcards, however, does not stop here.

Berlin, 2002. While waiting in the hotel foyer, I was skimming through one of those stands of free postcards that can now be found everywhere. Amongst the various adverts for films, cars, cabarets and so on, I suddenly noticed a puzzling card which, unlike the others, did not clearly show what it was for: there was a blue background, and a very large close-up of what I took to be a finger squeezing a tube of white and red toothpaste, out of which is coming, not the toothpaste that we expect, but a green caterpillar landing on a translucent, yellow toothbrush. It was only by looking at the bottom right of the card that I found the caption that explained all: *Jewish Museum of Berlin: not what you expect.* The association was easy: the green caterpillar, a revolting creature, represented Nazism (the green uniforms of the Wehrmacht); and the meaning was clear: one day you will be brushing your teeth and find that the beast is within you. Obviously, I can't tell whether that was what the card's designers wanted me to think, but even today I cannot dissociate myself from this reading. My reaction was no less immediate: this card is over the top. Not only did I consider the card to be in the worst possible taste, but I was shocked by this manner of advertising the Jewish museum.

Later I found out that this card was one of a series for the Museum in which all the cards work on the same principle. There is one of someone about to wash their hands; they put one hand under the tap from which not water but a sort of paste, a white-ish glue is running. The moral is: If I wash my hands of it, this is what will happen: I will be stuck, glued, and this will happen to me in my home, in my daily routine. Another card uses an even more disturbing idea: a high-angle close up of grass, a section of a felled tree trunk: the section looks like a slice of *saucisson*. Looking at this image, I cannot avoid making a scandalous association: the Jews (the tree) are like the flesh, the meat of the *saucisson*. Going beyond the symbolic towards the 'semiotic' – the semiotic functioning at the level of our drives, and urges (Kristeva) – this photomontage creates in me a psychosomatic effect: a feeling akin to nausea. Thus, like the optical illusion cards above, but in a more aggressive manner, the series of postcards for the Jewish museum make me feel ill at ease.

We know that the Jewish Museum itself devised this strategy. Much has already been written about this controversial building by the architect Daniel Libeskind. Even though I had steeled myself by looking at photographs before coming to Berlin, the outside of the museum was still a huge shock, all the greater for the contrast with the baroque-style building (the former Historical Museum of Berlin, designed by Heinrich von Gerlach) of which the Jewish museum is an extension, or rather a hideous excrescence: extremely high concrete walls lacerated in random fashion by a number of gashes form a line broken at irregular intervals. My cards were telling the truth: this was not what I was expecting. The entrance to the museum is no less surprising; in fact, the museum does not have an entrance to speak of: I enter via the baroque edifice with its Biedermeier décor, and then find myself in an immense lobby from which stairs descend to the floor below. The museum's entrance proper is thus underground. There, the space tightens to narrow corridors where the floor is not level, but on a slight slope. One of the corridors leads to the Holocaust tower and when the heavy door closes behind me I am as if crushed by the height of the bare concrete walls. Unheated, unlit, apart from a few angular openings, the space produces a strong effect of claustrophobia and of separation from the world; for a few moments I feel as though I will never be able to leave this room, and I am truly relieved when finally I push open the door to my liberation. Another corridor leads to a garden composed of columns filled with earth spaced apart at equal distances; out of them greenery grows and when I walk between the columns, I lose my balance, because the ground is sloped whereas the columns are most definitely vertical. We physically experience the destabilization of exile.

Everyone says that the Wall has been made to disappear but, walking around Berlin streets, I discovered the complete opposite: the Wall is everywhere. Even where it did not use to be, pieces of it exist to remind us. But Berlin has also created spaces of memory where the Wall is deployed more fully. This is the case of the East Side Gallery where a portion of the Wall has been given over to artists who have decorated it. Less well known but more moving still is the Wall Memorial on Bernauer Strasse, one of the mythical sites of Wall's history, in particular because many spectacular escape attempts were made at this point. The site is also famous for the destruction of the neo-gothic Church of the Reconciliation, photos of which went around the world (because of the Wall, the church found itself right in the middle of no-man's land); finally, this place is famous because the Wall at this point juxtaposes the Sophien Parish cemetery, home to the tombs of famous artists, some of which were moved at the time the Wall was erected.

The memorial takes the form of two big, brushed steel plates enclosing a 70-metre section of the Wall from which the graffiti has been carefully removed. The Wall is thus reflected to infinity in the surfaces of the sheets that face each other. But the full force of the installation lies in the wooden fence, which not only prevents us from accessing the space behind the wall, but which prevents us from even seeing what there is between the two enclosures. If I really want to try and see what is in this space I have to twist my body to look between the gaps situated at intervals between the slats of the fence (Wall pilgrims on occasions slip flowers into these spaces). Obviously, having made this physical effort, I discover that there is nothing to see. The space behind the Wall has been left as it was when the Wall was in operation: abandoned. It is a dead space. In the space of a moment, I am brought to the share the same feelings of frustration that the Wall provoked in those whom it separated.

The Wall Memorial, the Jewish museum, do not stop at conveying a message, or commemorating a past event; they try to make us physically experience the unease of Germans today confronted by their past. It is striking to see how this approach seems to have been taken up by the people of Berlin as form of necessity. We can, for example, wonder why Berlin has felt it necessary to build a Holocaust Memorial when the concentration camps are still there. But there is more. At the beginning, the Holocaust Memorial was a project conceived by a team of historians whose mission was to create a 'place for information', but the final product, *Fields of Memory´* by Peter Eisenman does not work at all like that. Two thousand seven hundred stelae are spread out over a surface area of 20,000 square metres, designed in such a way that the visitor, having entered into the area, will never see its outside edges. Between the concrete plinths, it is deliberately too narrow for more than one person at a time. Moreover, the way that the plinths are laid out, in waves of varying widths and heights, is designed to destabilize the visitor in so far as their perception of the whole will vary from instant to instant. According to the architect, the purpose is to reduce to zero the illusion of security.

The fact is that however loudly Berlin proclaims its postmodernity, its identity as a city-in-the-making, entirely open to business and the future, its uncomplicated Americanization – 'Potzdamer Platz or 'America Light', as Margaret Manale has it (2003, p. 209) – it is still the case that this image is doubly flawed; what Régine Robin calls the double 'ghostly machinery of German memory' (2001, p. 93): the Jews, and the Wall, which prevent Berlin from rebuilding its identity. 'The Germans, wrote Günther Grass, have lost their identity. They are missing things that they cannot recover, however hard they try: there is a hole in their consciousness' (1998, p. 253). Under these conditions, we can hypothesize that cities are like people in that, following a trauma, they need therapy.

Berlin has undertaken its own psychotherapy by taking the whole world as mediator, from whence a strategy, consisting of producing 'installations' which create physical unease amongst those who visit or look at them. In this way, Berlin is part of what I would call the *paradigm of intimate exposure*. The parallel here with 'reality TV' is striking: Berlin is exposing its intimate secrets to the public because Berlin cannot heal itself alone. Of course, in this case, the intimacy is collective, not individual, and of course we are entitled to think that there is no comparison between Berlin's problems and the trivial personal issues exposed for all to see on the television, but the process is indeed the same: the invoking of a mass third person, here humanity in its entirety, to solve one's personal problems and reconstruct one's identity.

Conclusion

These three analyses reach the same conclusion: the cities in question, each in its own way, demonstrate the need to open themselves to the outside world in order to forge a new identity. The progress invariably begins with a gesture of refusal: the refusal of an identity defined in terms of an ethnic collectivity (Bilbao), a traditional collectivity (Palermo), a national collectivity (Berlin). In order to build themselves a new identity, our three towns have had recourse to art, art conceived as media: Bilbao has made of art the definitional media of its identity, transforming the city into a sort of gigantic art gallery; Palermo has called on abstract sculpture and photography for the values transported by these art forms; Berlin has used urban art (museums, memorials) as a psychological operator. This manner of seeing art in terms of its public (or political) ends has often been denounced, especially by Hannah Arendt (1972). More recently, Henri-Pierre Jeudy (1999) has stigmatized the reductionism by which art is seen to have an 'operational social function', and the contemporary artist the function of 'mediator'. I believe that art and artists have never successfully escaped this functionalism as mediators, but will not go into the subject any further, since the issue here is not to know whether it is good or bad for art to take sides in questions of identity (we know that it does in the case of our three cities), but to understand why these cities took the decision, regarding their identity, to have recourse to art rather than to other media.

My hypothesis is that in a crisis situation such as that experienced by our three cities, traditional discursive media (the press, radio, advertising) are no longer sufficient, not even the television, since it has been compromised by spectacle, and thus that there is a need to have recourse to media that have a stronger *aura*, media that are capable of reaching everyone in their deepest soul. From this perspective, the choice of art as medium seems pertinent, although we need to be clear about the manner in which art is invoked in these cases: less for its representative or even symbolic function than for its *phenomenological* function; its capacity to stimulate first-hand experiences which involve and possibly disturb us. Bilbao invites us to share in its experience of heterogeneity and a learned plurality; Palermo brings us face to face with the experience of death; and Berlin plunges us into the experience of shared unease. In each case, the hope is that these experiences will lead to awareness, but also to a change of behaviour and of our vision of the world.

This manner of putting emotion at the heart of the chosen communication strategy is not without its dangers. Thus in the case of the 'compassion' of Berlin, we evoked the danger that visitors would not understand but feel, empathize. Then there is the risk of manipulation: the visitor is forced to undergo a certain impression rather than working on their own perception and powers of thought. To be certain, these fears are not without foundation, but it is also reasonable to be of the opinion that in order to escape from an identity trap, the call on emotion is unavoidable. As Herman Parret notes, 'politics is the intermingling of the social and the sensitive – politics is the dynamics of *sensus communis* (where the social is sensitised and the sensitive are socialised)' (1999, p. 224). From this perspective, we have to recognize for our cities the perfect fit between the identity rebuilding process and the way in which media-art has been used: Bilbao, faced with the risk of territorial enclosure, has used baroque art for its very vitality, its openness to a borderless community; Palermo, faced with its enclosure within a Mafia 'culture' of cultivated opacity, has had recourse to contemporary abstract culture and to photography for their evidential quality and their capacity to open a window onto a community of ethics; Berlin, finally, faced with its enclosure in an image that it finds hard to bear (the German identity reduced to the Jewish problem and the question of the Wall), has turned to urban art installations capable of conjuring up the humanity that is in each and every one of us.

One final remark. It might appear as if all these efforts regarding identity make no reference to Europe. However, this picture needs to be modified somewhat: the work on Bilbao's identity is part of the great tradition of European cultural cosmopolitanism; in Palermo, the defence of human life and human values not only breaks with the system of traditional Mafia 'values' but, equally, as the reference to the death row occupant Joseph O'Dell illustrated, with what is happening in the United States (Palermo against the death penalty). Berlin, finally, has focused the reconstruction of its identity around the idea of humanity as it is defined by the European philosophical tradition, as the universal collectivity of human beings. Thus Europe is indeed present, at the heart of the processes in operation in our three cities. Of course, each city tells its own story, but within a large 'framework' that they share, implicitly: a Europe of values which contrast with the dominant discourse of the loss of identity caused by generalized Americanization, globalization and the dominance of the laws of the market. Ultimately, the solutions devised by each of our three cities to their crises of identity give us reasons not for despair, but for hope.

Translated by Helen Drake

References

Abdallah-Pretceille M. & Porcher L. (1996) *Éducation et communication interculturelle*, Paris, PUF.
Arasse D. (1996) *Le détail. Pour une histoire rapprochée de la peinture*, Paris: Flammarion.
Arendt H. (1972) *La crise de la culture*, Paris: Gallimard-Folio.
Barthes R. (1980) *La Chambre Claire. Note sur la photographie,* Paris: Cahiers du Cinéma, Gallimard, Seuil.
Battaglia L. (1999) *Passion, Justice, Liberté*, édition Melissa Harris, Actes Sud/Motta, (in English, New York: Aperture, 1999).
Damish H. (1972) *Théorie du nuage*, Paris: Seuil.
Grass G. (1998) *L'Allemagne et sa mémoire*, Paris: Odile Jacob.
Jeudy H-P. (1999) *Les usages sociaux de l'art*, Paris: Circé.
Ladd B. (1997) *Ghosts of Berlin*, Chicago: University of Chicago Press.
Manale M. (2003) 'La modernité fait mythe' *Les Temps modernes*, no. 625.
Mons A. (2002) *La traversée du visible. Images et lieux du contemporain*, Paris: éditions la Passion.
Parret H. (1999) *L'esthétique de la communication. L'au-delà de la pragmatique*, Bruxelles: Ousia.
Ponge F. (2002) 'L'Ecrit Beaubourg', *Œuvres complètes*, vol. 2. Paris: Pléiade.
Robin R. (2001) *Berlin chantiers. Essai sur les passés fragiles*, Paris: Stock.
Tissander, Denis and Jean Modot (2000) *Sicile*, Paris: La Manufacture.

GETTING AWAY AND GOING HOME: THE VISUAL EXPERIENCE OF THE EXILE AS TOURIST

Karin Becker

A few years ago I was on a flight from Stockholm to Prague, taking advantage of inexpensive plane fares to spend a long weekend with a friend in a city where we had never been before. Once the plane was in the air, I noticed that many people got out of their seats to walk up and down the aisle, greeting and chatting with other passengers in what I assumed was Czech. They seemed to know at least half the people on the plane! In the meantime, other passengers, mostly women around my age, remained in their seats, reading through tourist guidebooks and talking with their seatmates. There appeared to be two distinct kinds of tourists on this flight: those like myself who were looking forward to exploring a new city, and those who were returning to a place where they once had lived. The man sitting next to me confirmed my observation. A documentary film-maker employed by Swedish television, he had grown up in Prague, and had left as a young man to make his home in Sweden. After many years away, he was now able to return regularly to visit friends and relatives, and occasionally on a professional assignment. In the course of our conversation, he offered me many tips about how to get around in the city and what to see, demonstrating both his personal familiarity with the place, what he missed and liked to do when he came back, and also his knowledge of what a visitor should know in order to negotiate the city as a tourist. The experience represented by this former Czech exile and his fellow passengers became the point of departure for this study, an examination of what it means to return as a tourist to a place once regarded as home after many years of being away; in exile.

It is evident that the political and economic changes that have reopened many European borders enable many exiles to return to the place that once was home. For them, memory of home is mixed with pictures and accounts, both private and in the media, that they have had access to over the years. What identities are actualized and come into conflict in the experience of returning to visit a once familiar place?

The contemporary European context is particularly significant in this regard, and often overlooked in migration studies. The turbulent political history of twentieth century Europe includes a complex range of events that in one way or another have had an impact on virtually every individual on the European continent. The re-drawing of national and political boundaries during and following two world wars, the Holocaust experience, the division of Germany, the uprisings in Hungary and Czechoslovakia and the final collapse of the eastern Bloc have all given rise to movements of people within and across old and newly-drawn borders. No person living in Europe today remains unaffected by this history, and the many who have moved – or returned – in response to one or more of these cataclysmic events carry the personal memories of its impact on their lives. The possibility of returning to a place one was forced to leave actualizes the complex process of negotiating the relationships surrounding the concept of home with all it represents, and the question of how one constructs identity with reference to place.

I have chosen to focus on how the returning exiles' experiences and identities are represented in the private photographs that are made and kept by family members. I look at the exile/tourists' photographs and films of the place they grew up, tracing changes in the visual documentation from before the period of exile through the photographs and films made during return visits as a tourist. How does the process of negotiating identities around the complex experience of migration and return make use of private images and narratives as well as other media forms? My purpose is twofold: first, to explore how visual documents and the narratives they give rise to are used in identity construction, and second, to use this process to gain insight into the complexity of cultural identities being created in contemporary Europe. Does the negotiation of identity include references to sense of self as pan-European? Can the identity of the former exile accommodate the notion of becoming a citizen of Europe?

Visual media and returning 'home'

This is a study of media production, understood here as the tourist/exile's own visual media practice, including how these documents are edited and framed in re-presentations that form a narrative of personal history, memory and identity. The private use of photography must further be seen as a significant form of *mass* media production (Becker, 2002). As Benjamin so richly describes, at the turn of the nineteenth century in Paris photography became inscribed in the culture of consumption at the same time that it was used to expand the visual marketing of consumer goods (Benjamin, 1982/1999).[1] With the family and the tourist as primary targets, a band was established between photography, leisure, and the visual documentation of experience that has survived well into the twenty-first century (Slater, 1995). Despite vast changes in photographic technology, the camera continues to be a naturalized attribute of the tourist. And the tourist continues to be bombarded by images of the sites to be visited, in travel posters and brochures, on postcards, and in the range of television programmes and websites directed at the would-be tourist. For many exiles, the mass-mediated image of 'home' is also found in documentary and news formats, as a place ravaged by conflict or poverty or both.[2] These are in addition to the private photographs that the exile/tourist has of the people, events and environments that represent the place of origin.

I have argued for using the word 'vernacular' to encompass the range of photographic practices people carry out in their private, everyday lives (Becker, 2002). To call this 'amateur photography' excludes the ways professionals use photography in the private sphere. Many of the former exiles I encountered in this study worked with visual media in a professional capacity. Yet they had also taken photographs of family and friends, often using formats that differed distinctively from their professional work. The other common term is of course 'family photography'. This, however, excludes many of the subjects and occasions where family members are absent from the images. Further, when the terms 'family' and 'photography' are linked, a mutual construction is established that elides the dynamics and exercise of power that take place within each.[3]

The term 'vernacular' captures the common and regular ways photography is integrated into everyday life and memory. Vernacular refers to forms of cultural expression arising through everyday use in specific historical, social and cultural constellations. It has therefore the advantage of describing cultural forms that draw from a broad spectrum of practices, and from a range of forms and expressions. Thus, vernacular photography can be seen as a visual dialect that incorporates elements from other visual genres, from popular culture and other media that cross its path. Interwoven with and distributed through a variety of media, the vernacular is clearly recognizable as a transnational phenomenon (Becker, 2002; Crang, 1999; Crouch and Lübbren, 2003; Naficy 1999; Urry, 2002). The concept of the vernacular encompasses the 'ordinariness' of these photographic practices, in the sense that they are grounded in what is easily recognizable as *alltagskultur* or the *quotidien*.[4] It should be understood as also including the occasions when the camera is used to document that which is precisely *not* part of the everyday, such as reunions, holidays and other celebrations, where photography is self-evident in the ritualization of the event. Referring to this as a vernacular form draws attention to photography's relation to the range of activities that form the stuff out of which people interpret and assign meaning to their experience.

I do not, however, wish to privilege these forms of visual documentation as exclusive bearers of meaning in the reconstruction of memory and identity. To re-experience the language of her childhood can bring about profound reinterpretations of the exile's personal history (Schüllerqvist, 2005). Music and smells can also reawaken once-lost memories, making them available to weave into narratives of identity. Yet visual documentation is, arguably, the most common way of recapturing the places, people and memories of the past and reintegrating them into one's ongoing life. For the exile, it can also provide a bridge between the new and old culture, as witnessed by the many emigrants who use visual media in their professions. Several of the people I encountered in this study, like the man I met on the plane to Prague, work as photographers or film-makers, and use their professional skills to create opportunities to travel between their two 'home' countries.

Vernacular photographs are also used to renegotiate cross-generational identities as families move or are cut off from familiar people and places by the redrawing of political borders. This is suggested by interviews Meinhof and Galisinski conducted with people from different generations living in a town on the border between the former GDR and West Germany (Meinhof and Galisinski, 2000). Integrating photographs of the town into the interviews gave rise to narratives that revealed the complexity of constructing an identity in relation to radical changes in the socio-political environment.

On a personal note: for me, as a Swedish-American, re-discovering my mother's photographs and diaries from her first visit to her cousins in Hälsingland, had a profound effect on how I explained to myself my own move to Sweden as an emigration back to my grandparents' 'homeland'. In my own photographs from early visits I see efforts to find visual continuities between my Swedish heritage and my somewhat uncertain search for a new European identity for myself. Today, twenty years later, I trace my dual citizenship back through an ambivalent quest for what it meant to be both a tourist and an immigrant, the returning child of Swedish emigrants.

Consider how much more complex that quest is for people whose histories of crossing national borders are written as forced movements, escapes, and restrictions governing where they may settle and make themselves a home. How is this complexity reflected in their relationship to the land that once was home? What is the process of re-negotiating one's own relationship to the place one was forced to leave, and to the people who remained there? Much recent work on migration has focused on the complexities of returning, pointing out that few people emigrate for good. As anthropologists Lynellyn Long and Ellen Oxfeld point out (2004), return is an integral part of the migration experience. While responding to larger universal processes (globalization, transnational movements and processes), returns reflect particular historical, social and personal contexts. Specifically, looking at returns 'allows us to analyse these larger processes in terms of people's own systems of meaning and experiences and to discern the particular human consequences of these larger forces in everyday lives and actions' (Long and Oxfeld, 2004, p. 3). What part do photographs and *photographing* (the act itself being important to the process) play as one negotiates one's identity around the unstable concept of home?

Between two homes

The material I have gathered is primarily from people who spent their childhood in countries which belonged to the Soviet block, and who went into exile in 'the West'. At some point, depending on which country they had emigrated from, it became possible for them to return 'home' for fairly regular visits. Asking them to show me their photographs, and interviewing them about what these images reveal to them has indicated a series of stages in a process of how one locates or constructs a hybrid identity, rooted in two places, the country of origin and the new home (in all cases this has been Sweden).

For the purposes of this article, I focus primarily on one family whose experience reflects on a personal level many of the historical events that have redrawn European political boundaries. They emigrated from Hungary to Sweden in 1956, and made their first return visit in 1963. They were Jews, Holocaust survivors who had remained in Budapest. They were also socialists, and initially viewed the Soviet Union as liberators. This dream destroyed, the couple decided to leave with their five-year-old daughter during the tumultuous weeks of the Hungarian revolution. They settled in Stockholm and stayed in touch with family members who remained in Budapest. On return visits, they stayed in a hotel (always the same one) and spent their time with family members and old friends. Eventually, the daughter began to visit Hungary by herself, occasionally taking a friend with her, and made friends with young people her own age who were not family members. Today, the parents are no longer living, and the woman still lives in Stockholm, travelling occasionally to Hungary in her work as a theatre director.

This family's collection includes a variety of forms of visual documentation relevant for this study: first, a few photographs, several of them framed, taken in Budapest during and shortly after the war; second, a large number of photographs – both prints and slides – made by either the father or daughter during visits to Hungary from 1963 onward, and finally several super-8 films the daughter made as a teenager during subsequent visits.[5] Rather than attempting to look at the family's entire collection (which also included many films her father had made), I asked the daughter to make a selection that could represent the experience she and her parents had of moving to Sweden and their return visits to Hungary after years of being away. This interview, like others I conducted, followed a pattern of looking through the photographs, roughly in chronological order, as the daughter first described the family's background and history, and then used the photographs and films she had selected as a framework for presenting how she experienced the visits she made, initially accompanying her parents, and later on her own. This woman's account has clear parallels with those of other European emigrants who, having moved as children, integrate their parent's often traumatic and tragic histories with their own, as they construct narratives based in part on photographs from their past (cf. Spitzer,1999; Schlesinger, 2004). Like others I have interviewed, this woman's narrative follows a number of stages in a process of negotiating an identity for herself that takes account of the conditions and experiences of migration and return.

Pictures from before

The first stage revolves around photographs taken prior to immigration that carry meaning as aspects of a personal and collective past. The earliest photographs she chose to show me, from the war and early post-war period in Budapest, are valued as important links in her own history. The photographs provide a point of departure for the narratives told and retold over the years and that recount how her parents met in the heady days at the end of the war, the search for family and friends, the uncertainty over who had died and who had survived.

These photographs carry different meanings for different family members. For some, like this woman's father, photographs provide evidence that one is a survivor. According to his daughter, photographing was a way he established continuity between the various places he had lived. For others, like her mother, the photographs symbolize not continuity, but ruptures in the fabric of the past, showing how different her life was before the move to Sweden. The early photographs from Hungary helped to sustain a nostalgic bond to her past life, at the same time that they confirmed her alienated relationship to the culture of Sweden.

The daughter, in turn, accounts for her own ambivalent relationship to her country of origin by referring to the conflict between her parents' respective relationship to their past. In the photographs she finds evidence that confirms this inner conflict: she notes her father's proud bearing in photographs taken of him in uniform, at the same time that she describes how the Jews enlisted in the Hungarian army were placed in the front ranks; used as 'canon fodder'. She refers to the film *Life is Beautiful* (Begnini, 1997) to explain his rakish appearance; her father was like the father in Roberto Benigni's film, attempting to protect his wife and child from the harsh realities of the world by maintaining a positive and jovial attitude toward life. Her mother's stylish appearance, on the other hand, was not a denial, but an attempt to conceal the trauma of a second uprooting, first to the concentration camp at Terezin and then to Sweden. Even in the earliest photographs from the post-war period in Budapest, the daughter sees evidence of her mother's fragility, behind the flirtatious smile as she cocks her head for her husband's camera. As she described the different meanings these photographs have carried for different members of her family, she is revealing the complex negotiations that constitute these images and, in turn, family life (cf Hirsch, 1999, p. xi).

Early reunions and a widening gulf

A second phase or aspect of identity formation occurs around the photographs made during the first return. After seven years living in Sweden, this family made a trip to Budapest to visit family and friends. The pictures made during this trip show family members posing together for the camera. The contrast between them is stark, clear evidence of how the returning family has changed during their experience living in a western European country. The daughter, who was 12 years old at the time, finds the photographs taken of her together with her slightly younger cousin particularly meaningful. The cousin wears her hair in tight curls, and has a frilled blouse over a full skirt. The daughter has her straight hair pulled back into a simple ponytail and is wearing dark slacks and a simple sweater. She represents the style becoming popular among teenagers in western Europe. She also points out how stylish her mother is in the photographs compared to the other women in the family. Although always careful about her appearance, her mother's hair and clothing offer additional signs that they are now from 'the West'.

In two other pictures she is carrying the toy bear she left behind. She had been told to present this precious toy to her cousin when they left Budapest in 1956, and now it had been returned to her. Taking photographs of the cousins together, and together with the toy bear, was clearly intended to symbolize the reunification of the generation that had been separated by the migration experience. For the daughter, however, the photographs with her cousin rather show evidence of the rivalry between them that was to persist until they reached adulthood.

What is not documented is also significant, and often forms a darker part of the narrative of return. Behind the photographs of family members smiling for the camera is the account of the return itself, which is usually not documented. This narrative includes the work involved in organizing the correct papers, the route taken and the tension around border controls, all of which compound the anxiety about what it will be like to come back. In this family's case, the daughter remembers clearly her mother's acute anxiety during the long journey. They traveled by train, taking a circuitous route that entailed many changes. Her mother's nervousness was further heightened by the insults of the uniformed men at each checkpoint, as they examined the family's new Swedish passports and searched their luggage before allowing them to continue on toward Budapest. The narrative of the trip back is fused with longing and fear, none of which is evident in the photographs. Yet the images, which on the surface appear to follow the conventions for photographing family reunions – presenting various relatives side by side for the camera – unavoidably give rise to memories and accounts that compound the meanings attached to these highly ritualized photographic moments.

There are a number of themes in the photographs from the first return that are consistently repeated in subsequent visits. Yet none of these are what could typically be called 'tourist' pictures. There are no photographs of public places or the sights one sees on travel posters and in tourist brochures. Family gatherings, at home and on the beach, are the primary themes of the exiles' photographs of return visits. The accompanying narrative explains relationships among family members and friends, often with anecdotes that draw attention to a particular aspect of the image that might otherwise seem insignificant or even be ignored. Following this pattern, the daughter's narrative focuses on the family relationships, the beachhouse they usually visited, the close tie between her father and his older brother, the continuing jealousy between her and her cousin, and the ongoing ambivalence that accompanies these visits. These could be photographs from any family reunion, anywhere, replete with the subtext that only a family member can read.

At the same time, however, the narrative of the return weaves in numerous points of comparison between the returning family and those who remained behind. This ongoing comparison is the specific subtext that distinguishes the meanings of these family reunion photographs from others which would appear to be similar. The differences between the women's clothing style is one example of such a comparison in the case of this family. The daughter also points out home interiors that she sees as dark and heavy in contrast to Swedish styles of the same period. Various objects, such as the imposing stereo in her aunt's living room, are also points of pride for the Budapest family members. She stresses, with reference to the photographs, how important it was for her relatives to show the richness of their lives to those who had left.

Food often provides an important link to memory, particularly traditional dishes that are prepared with ingredients that may be hard to obtain in the new country and that are part of family rituals.

This family was no exception, and central to the daughters' memories of returning to Budapest are the many family dinners, with traditional dishes in a quantity that she eventually finds repulsive. She showed several photographs without any people in them, showing only a table with a lace tablecloth and laden with an overwhelming array of different dishes. These images serve as symbols of rich cultural traditions that are hard to sustain for those who moved away. Food itself becomes a symbol of what the relatives offer to their guests, and the daughter recalls how they were urged to eat, and her awareness that she hurt the feelings of her hostess when she protests that she can eat no more. In the ambivalent memories these photographs arouse she locates the origins of her acute loss of appetite, a problem verging on an eating disorder.

Creating new reunions

At some point during the exile's repeated returns, photographs begin to depict scenes and places not directly associated with visits to family and friends. In this family's documentation, a turning point occurs when the daughter starts to photograph on her own. After several years of regular visits to Budapest, she begins to spend time by herself in the city, visiting tourist sites that she has seen in the travel brochures, and photographing the people she sees. For the first time in this documentation, we see what might be called tourist pictures. Looking at these photographs thirty years later, she interprets them as part of her effort to break free of the family and come to her own terms with the city where she was born. In many cases she can recall the feeling she had taking a particular photograph. One, for example, shows two ordinary-looking women on a streetcar, apparently unaware that they are being photographed. She recalls how pleased she was that she had captured this everyday scene from life in Budapest. She also took self-portraits, seating herself at the foot of statues of the Hungarian poets and authors she has read. These photographs, in particular, show the process of constructing an identity that includes a cultural heritage not available in her Swedish life. During this period, she also returns to the first place she lived and photographs the entrance and the courtyard of the apartment compound, clearly as a way of recovering and incorporating her past into the person she was becoming.

A social aspect also enters some of the photographs, and is further evident in the super-8 films she begins to make around the same time. A group of young people appears; she now has her own friends with whom she gets together on visits to Budapest. She met several of these young people through her cousin and has now, as she explains, taken over her cousin's group of friends. Many of the pictures portray scenes typical of city visits: friends in a park or in front of a city monument or building. Some of the photographs and film sequences are playfully staged performances for the camera. In one she is sitting in front of a pile of empty plates with a mock grimace on her face. She notes how extremely thin she looks, and recounts that her friends are playfully urging her to eat more.

In this phase, photography itself has become a way of establishing a reflexive relationship to the former home. The exile is clearly not free of the memories of the past, but is exploring ways to integrate them into a new identity. Here, the camera becomes a tool in an active process of negotiating between a past over which one had only minimal control, and an identity that is constructed out of selected aspects of that past, and links them to the person one is becoming. The photographs and films made during this period of the emigrant's life are part of a narrative that bridges the moves that are continually being drawn between the old and new home, as she works to integrate them into a whole that makes sense to her.

Pictures and memory as creative resource

There is, finally, a way that exiles document their returns that can also be observed in their selection or reinterpretation of earlier photographs and which points forward in time. Specific objects, people and themes are seen to recur in the earlier photographs, and eventually form narrative threads that tie together experiences from the past with the exile's future. The way a landscape looks, a particular kind of flower, houses painted in a particular colour are examples of phenomena can exert a pull on the exile, and that are found to carry meaning when they then recur in the photographs taken of the former home. Discovering one of these signs repeated in several photographs is often experienced as an epiphany; it explains to the exile an aspect of who he or she is. Among the people I interviewed who work in the visual arts, these threads from the past often became themes for projects that they chose to develop professionally. For example, several have traced the fact that they continue to be drawn to documenting the lives of individuals on the margins of society to their own experiences of living between two cultures.

The woman whose photograph collection I have been discussing here also pointed to certain photographs and film sequences as sources she will use in her writing or other creative projects she is developing. The recurring image of the East German Trabants on the streets of Budapest is one example that she would like to work into a future project. She loves the boxy form of these old cars, unchanged from the early 1960s, and the memories they evoke. She recalls that her uncle had one, of which he was very proud. That car's final appearance is in the last photograph her father made of his brother, quite ill at the time, shortly before his death.

In this stage, the fact that the individual is able to move between the two cultures is seen as an advantage. The freedom to transgress the once-closed border in one's private and professional life (or lives) provides a way to weave together the multiple identities the exile has experienced. As the daughter in this family reported in our first interview, she still catches herself talking about Budapest as 'home', but always with the sense that she is placing quotation marks around the word.

The place of images in narratives of return

In addition to the various stages of migration and return that tie this woman's narrative to the broader cultural and political environment of post-war Europe, there are a number of recurring themes and characteristics that address the more general question of how private photographs and other media are used in the process of negotiating identity, particularly around returns. In planning for or imagining a return, photographs and other objects may be used 'to memorialize the past and to develop a symbolic architecture for a future return' (Oxfeld and Long, 2004, p. 8). For the exile unwilling or unable to make the journey, the photograph can provide a replacement, when friends, acting as surrogates, are asked to photograph once familiar places (Slyomovics, 1998, p. 16).

From anthropological studies of migration we also know that the experience of the returning migrant is seldom without problems (Long and Oxfeld, 2004). In the interviews I conducted, photographs and films are often presented as containing evidence of conflict between the emigrants and those who stayed behind. The woman I focus on here showed many examples where she could point to conflict that would not necessarily be visible to an outsider. She sees evidence of this and the pain it could cause her and her parents, for example, in the contrast between how her mother is dressed and the appearance of other women in the family, in photographs of her and her cousin together; in pictures of home interiors and of generous spreads of food. And, as her narrative unfolds in response to these images, other feelings of difference and estrangement come to light. She describes, for example, Hungarians' guarded criticism of those who left the country in 1956, which she encountered among both family members and others each time she returned.[6] Photographs, while not visualizing these experiences, can nonetheless provide keys that open them up for examination and reflection. This again is in line with research that identifies how returnees may simultaneously identify with and be estranged from former homelands, and particularly the people they are or have been closest to (Long and Oxfeld, 2004).

This also points to, as alluded to earlier, the limits of personal visual documentation. The most traumatic aspects of the emigrant experience are rarely photographed. The long account of this family's first trip back to Hungary, the circuitous route and all the border controls, replete with the insulting stares and comments of the soldiers, which left her mother a nervous wreck, remains one of this woman's strongest memories, although there are no pictures of the trip. In many cases, the visual documentation and verbal narrative are complemented and filled out with other media references that deepen the 'memory work' (Kuhn, 1995). The documentation of the Holocaust remains a fond that lies implicitly behind any account of a Jewish family's history and experience (Hirsch, 1997). Popular culture is also replete with such references. This woman's referring to her father as similar to the man in the film *Life is Beautiful* is an example of how intermediality can intersect with and give additional meaning to personal memory.

In other ways, however, photographic documentation *exceeds* the visual. Photographs as material objects carry significance that is frequently overlooked when focusing on what the image portrays, what it is of. Elizabeth Edwards has pointed out the importance of touch and gesture in the often unspoken relations people have with photographs that can in turn be linked back to the 'nature of the photograph and its indexical quality' (Edwards, 2005, p. 41). These aspects emerge as people show and talk about photographs, in performative interactions that fuse the image and its materiality and, Edwards argues, give sensory and embodied access to photographs. This is obvious in the ways people handle and present photographs in interviews, and use gestures and expressions to mimic, complement and contradict the visual. This became apparent in the course of the first interview with this woman whose photographs I discuss here. As she laid out the photographs on the table in front of us, commenting on each one, I gradually realized that she was placing them in particular relationships to each other, turning them over and making them overlap to form an increasingly intricate patchwork pattern. Through the course of the interview additional layers were added to the pile, a physical expression of the family history she was narrating. When I asked her about what was taking place, she confirmed that it was a conscious performance, that the photographs she had selected from the first trip back to Budapest were spread as a physical base upon which the other images would rest and build. She was clearly demonstrating how the photographs through their materiality become 'relational objects', referring to subjectivities and emotional registers that cannot be reduced to the visual apprehension of the image (Edwards, 2005; Edwards, Gosden and Phillips, 2006).

The photograph's kinetic and tactile qualities may in fact be more important than the visual as keys to memory. The indexicality of the photograph as object, that is, the sense that it bears the physical trace of its referent, may be more central to how we apprehend a photograph than what we see when we look at it. To support this argument, I return to the first photographs this woman made with her own camera, moving through the city by herself. The photographs she showed me from these journeys were the only ones that resemble what could be called tourist photographs, when she was using photograph to create a new relationship to this place of history, and to establish her own position within it. In order to accomplish this, she distanced herself from her family and adopted the attitude of the tourist, seeking out the landmark sites and photographing people on the tram (the city's 'natives') she did not know. As Marita Sturken has said, 'The tourist embodies a detached and innocent pose' in order to use the visual, the infamous 'tourist gaze', as the mode of contact with his surroundings (Sturken, 2006, p. 9; Urry, 2002). Photographing helped her to accomplish this distance, enabling her to create a liminal position that she could occupy while renegotiating her own relationship to the city where she was born. Following this period of distanciation and reflection, she began to reintegrate the memories of her family's past and her childhood into her own experience, establishing a hybrid identity that she feels accurately represents her past as part of who she is. The photographs she has from her family are again important physical reminders of who she is and where she comes from, woven into a narrative of memory and experience. Barbie Zelizer has developed the concept of 'retrospective renominalization,' a renaming of earlier events or places in accordance with subsequent events (Zeilzer, 1995, p. 222). In a slightly different sense, this 'naming' of the past also occurs when a person applies new significance to previous places and events, as a way of integrating them into what otherwise could be experienced as a bifurcated identity. As this Swedish-Hungarian Jewish woman now a successful director travelling regularly between Budapest and Stockholm in her work, exclaimed: 'Can you believe it, I was born on Raoul Wallenberg street in Budapest, right behind Vig színház, the Comedy theatre!' Long and Oxfeld (2004, p.6) note: 'People negotiate the multiple spaces and movements in forming complex and at times, seemingly contradictory identities. The repeated and circular migrations that are part of many returning immigrants' experience reinforce cultural hybridity'. Photography can serve an important role, as we have seen in this woman's life history, through a narrative performative process that fills it out and gives it added depth and meaning.

Becoming European?

I have relied heavily on one individual's narrative, based on a presentation of her family's photographs, in order to investigate how private vernacular images come into play in the process of negotiating identities around the complex experience of migration and return in the context of contemporary Europe. Can this case, together with the others I have examined, support the additional claim that a *European* identity is being constructed through this process? With regard to the specific and conflicted political history of Europe's redrawn boundaries and the movement of populations that have ensued, the answer must be yes. No private collection viewed or, as I argue here, *performed* in light of history can exclude references to the effect of that history on family members' lives and experience, regardless of the specific national context where the family made its 'home'. The people represented in this study were never *not* Europeans. Unlike many other contexts of forced migration, there was no delineation of authenticity that deprived them of a claim that they were European.[7]

On the other hand, adopting or negotiating a European identity cannot solve the problem of where one feels most at home. As soon as one moved, or left, a mutual process of distinction was put in motion, setting the emigrant apart from those who stayed behind. With the adoption of citizenship in the new homeland, the difference became even more firmly established, as the family at the centre of this study experienced. They became Other. In many cases it was only with the adoption of citizenship in a new country that one could return legally for visits with remaining family and friends. This fact does little, however, to mitigate the quality of otherness. Today, the freedom of movement bestowed on the holder of an EU passport only deepens the sense of difference between those who emigrated and those who stayed behind.

May Joseph discusses citizenship as performance, an expressive act, and introduces a distinction between legal and cultural citizenship as 'a structural means of locating specific currents, influences, and laws that affect people in distinct ways as migrant subjects' (Joseph, 1999, p. 5). Although she is concerned primarily with the postcolonial context through which contemporary East Indians migrate, the distinction is useful for thinking about the patterns of migration and return established by the Europeans I have interviewed. For many of them mobility between two or more locations is an important mark of who they are. The relative freedom of movement afforded by legal citizenship gives them the means to practice the forms of cultural citizenship that they have achieved and express in the often hybridic forms of their life and work. They live and work in Europe, as Europeans. At the same time, being able to return to their country of origin is important to their cultural identity. It is a resource they continue to draw upon, and they continue to refer to it as 'home' – safely framed and set apart by quotation marks.

Notes

1. Despite the interest in photography that runs through much of his work, Benjamin devoted remarkably little attention to the amateur photographer, and then in largely negative terms.
2. The film *The Terminal* (dir. Steven Spielberg, 2004, DreamWorks Distribution LLC) shows this dilemma in the character Viktor Navorski (played by Tom Hanks), from the country of Krakozhia, who lands in New York as a tourist, yet suddenly becomes an exile when a violent coup eliminates the possibility of his returning home. Trapped in the airport, he runs between TV monitors showing news footage of the coup, as he struggles to understand the English reports that would explain the terrible footage he is seeing.
3. Slater (1995, pp.129, 135). For other examples of work that deconstructs the image of an idealized family life created through these cultural practices see Barthes (1969), Kuhn (1995) and Hirsch (1997).

4. For discussion of meanings of the term 'ordinary' in cultural research, see Williams (1983: 225–27) and Couldry (2000: 45–46).
5. The work of artist and documentary film-maker Peter Forgacs is an obvious inspiration for this research, although the material from the families I interviewed are not part of the Private Film & Photo Archives Foundation in Budapest. The archives house a collection of amateur films and photographs spanning twentieth century Hungary. Another important source has been Roger Odin's research on the esthetics of amateur film (Odin, 2004).
6. Her description is consistent with the withdrawal into private life that characterized the cultural climate in 1970s' Hungary: 'Politics is a vanishing topic of conversation; after one or two jokes or anecdotes about the latest corruption scandal the talk turns to more important subjects, like one's summer house, vacation travel plans, matters of personal and domestic fashion, last night's TV movie, and marital problems' (Konrád and Szelényi, 1979:, p. 200; see also James, 2005).
7. May Joseph's examination of the experience of Asian-Africans during the 1960s and 1970s is a case in point, when 'the kinds of citizenship available to non-Africans in [their] former homelands' was sharply curtailed (Joseph, 1999).

References

Barthes, Roland (1969). *Mythologies*. New York: Hill & Wang.

Becker, Karin (2002). Fotografier. Lagrade bildminnen, in Karin Becker, Erling Bjurström, Johan Fornäs & Hillevi Ganetz *Medier och människor i konsumtionsrummet*. Nora: Nya Doxa.

Benjamin, Walter (1982/1999): *The Arcades Project*, Cambridge MA/London UK: The Belknap Press of Harvard University Press.

Couldry, Nick (2000): *The Place of Media Power: Pilgrims and Witnesses of the Media Age*, London & New York: Routledge.

Crang, Mike (1999). Knowing, tourism and practices of vision, in Crouch, *leisure/ tourism geographies*.

Crouch, David (ed.) (1999). *leisure/tourism geographies. Practices and geographical knowledge*. London: Routledge.

Crouch, David, & Nina Lübbren (eds) (2003). *Visual culture and tourism*. Oxford: Berg Publishers.

Edwards, Elizabeth (2005). 'Photographs and the Sound of History.' *Visual Anthropology Review* 21(1 and 2): 27–46.

Edwards, Elizabeth, Ruth Phillips & Chris Golden (eds). (2006). *Sensible Objects: Colonialism, Museums and Material Culture*. Oxford: Berg.

Hirsch, Marianne (1997). *Family Frames. Photography, Narrative and Postmemory*. Cambridge: Harvard University Press.

Hirsch, Marianne (ed.) (1999). *The Familial Gaze*. Hanover: University Press of New England.

James, Beverly A. (2005) *Imagining Postcommunism: Visual Narratives of Hungary's 1956 Revolution*. College Station TX: Texas A&M University Press.

Joseph, May (1999). *Nomadic Identities. The Performance of Citizenship*. Minneapolis: University of Minnesota Press.

Konrád, György, & Iván Szelényi (1979). *The Intellectuals on the Road to Class Power*. New York: Harcourt Brace Jovanovich.

Kuhn, Annette (1995). *Family Secrets. Acts of Memory and Imagination*. London: Verso.

Long, Lynellyn D. & Ellen Oxfeld, eds. (2004). *Coming Home? Refugees, Migrants, and Those Who Stayed Behind*. Philadelphia: University of Pennsylvania Press.

Meinhof, Ulrike H. & Dariusz Galasinski (2000). Photography, memory and the construction of identities on the former East-West German border, *Discourse Studies*, Vol. 2:3, pp. 323-353.

Naficy, Hamid (ed.) (1999). *Home, Exile, Homeland. Film, Media and the Politics of Place*. New York: Routledge

Odin, Roger (2004). De beaux films de famille, in Valérie Nahoum - Grappe & Odile Vincent (eds.) *Cahiers d'ethnologie de la France,19, Le goût des belles choses*. Paris: Èditions de la Maison des sciences de l'homme.

Oxfeld, Ellen, & Lynellyn D. Long (2004). Introduction: An Ethnography of Return, in Lynellyn D. Long and Ellen Oxfeld, *Coming Home?*

Schlesinger, Philip. (2004). W. G. Sebald and the Condition of Exile, *Theory, Culture & Society* Vol. 21(2): 43-67.

Schüllerqvist, Lotta (2005). Resan från barndomens till nutidens Polen, *Dagens nyheter,* 9 August 2005, p. A 18-19.

Slater, Don (1995). Domestic Photography and Digital Culture, in Martin Lister, ed., *The Photographic Image in Digital Culture*.. London: Routledge**:** 129-146.

Slyomovics, Susan (1998). *The Object of Memory: Arab and Jew Narrate the Palestinian Village*. Philadelphia: University of Pennsylvania Press.

Spitzer, Leo (1999). The Album and the Crossing, in Marianne Hirsch, ed., *The Familial Gaze*, pp. 208-220.

Sturken, Marita (2006). Tourists of History: Souvenirs, Architecture, and the Kitschification of Memory, Keynote paper presented at the conference *Technologies of Memory in the Arts*, Radboud University, Nijmegen, the Netherlands, May 19-20.

Urry, John (2002). *The Tourist Gaze*. 2nd ed. London: Sage.

Williams, Raymond (1983). *Keywords*. 2nd ed. London: Fontana.

Zelizer, Barbie (1995). Reading the Past Against the Grain: The Shape of Memory Studies. *Critical Studies in Mass Communication* 12: 214-39.

Exiles and Ethnographers: An Essay

Philip Schlesinger

For Gertinko, whose photograph is all that's left.

I

In this essay I wish to reflect on some contemporary representations of the condition of exile, in particular the authorial desire to document bodily, psychic and spiritual dislocation in terms that are broadly ethnographic.

As my main interest lies in opening up a way of thinking about this terrain, I have concentrated principally on some writings by the German-born, British-domiciled novelist and academic, W. G. Sebald (1944–2001). For me at least, these writings have offered an important entry point into the questions addressed. Much of Sebald's work – I shall argue – is an outstanding example of how to undertake what elsewhere I have called a 'literary ethnography'.[1] This way of writing can be characterized as a mode of literary representation that underpins – and sometimes reflexively invokes – an *ethnographic* style, stance and methodology. It has a close affinity to contemporary anthropological ethnography that couples 'first-hand observation with interviews and with historical data and analysis of texts and imagery' (Macdonald, 2001, p. 72).

II

To play with the conventions that surround writing is the very stuff of much literary production. To reflect on how those conventions are used, and to demonstrate their effects, is not only of major concern to the literary critic but also of considerable interest for the sociologist of literature (Becker, 2001; Ruggiero, 2003). Some have taken the game to the very edge. Clifford Geertz

(1975, pp. 6–16) has famously suggested that ethnography is an intellectual project characterized by 'thick description', by its effort to interpret and analyse 'the structures of signification' encountered in the field. He has argued that anthropology produces *fictions*. He does not means that such accounts are therefore false but rather that they are fashioned out of layers of interpretation. It is not the mode of representation that matters, he maintains, so much as whether it offers us a cogent explanation of social action. While there are questions to be asked about Geertz's epistemology, for present purposes it is the explicit analogy that he draws between ethnography and novel-writing that is my key concern, for it provides an apposite entry-point to the argument.

W. G. Sebald's work has self-consciously situated itself on the borderlands of anthropology, journalism, the memoir and the travelogue. In Sebald's case, the generic crossings undertaken by the novelistic imagination involves a disruptive use of realism, intended to make us unsure about whether what we read is fiction or, say, autobiography. Raymond Williams (1976, p. 217) has noted that realism has had a controversial history 'as a term to describe a method or an attitude in art and literature – at first an exceptional accuracy of representation, later a commitment to describing real events and showing things as they really exist'.

The aspect of realist appeal of most interest here centres upon the use of *evidence* in literature – actually or ostensibly based in the gathering of documentation and testimony. The material amassed is open to diverse uses and forms of representation. Such gamesmanship derives from the inherent malleability of the codes that define distinct forms of writing. Often debatably, we may attempt to deploy criteria to distinguish between fiction and non-fiction, the literary and the non-literary.

These distinctions are themselves the outcome of a long social process, as Raymond Williams has further pointed out:

> the specialization of 'literature' to 'creative' or 'imaginative' works...is in part a major affirmative response, in the name of an essentially general human 'creativity', to the socially repressive and intellectually mechanical forms of a new social order: that of capitalism and especially industrial capitalism. (Williams, 1977, p. 50)

The Romantic notion of 'creativity' still underpins the conventional drawing of lines between fiction and non-fiction and connects directly to the employment of genres. Williams (1977, p. 185) sees genre as providing a 'new kind of constitutive evidence' of 'the practical and variable combination and even fusion of the social material process'. In short, a genre offers us a key to a wider understanding of how cultural products relate to social structures. But as Tzvetan Todorov (1990, pp. 10–11) has noted, 'literary genres...are nothing but...choices among discursive possibilities, choices that a given society has made conventional'. This is not to engage in thinking deterministically. The practice of writing involves making decisions about form and content, about where to place what is being produced, within limits that themselves are renegotiated over time.

Analogous practices *preparatory* to writing may lead to very different outcomes, depending on where you choose to situate yourself generically. The process of amassing various forms of empirical evidence by the historian, anthropologist or sociologist to produce, for instance, a canonically referenced monographic account, can also be used by a fiction-writer for writing a novel, by a film-maker producing a feature striving after authenticity – or indeed by a song-writer for writing songs. The precise ways in which the research is conducted may not be identical, as here too, we tend to follow conventions.

But the common denominator is the undertaking of *research*. Bob Dylan (2004, pp. 84–5), for one, has regaled us with his youthful reading lists, as well as his deep musical investigations, and has also recounted the time spent in that mythical repository, the New York Public Library, to prepare for painting his version of what he calls the 'American landscape':

> In one of the upstairs reading rooms I started reading articles in newspapers on microfilm from 1855 to about 1865 to see what daily life was like. I wasn't so much interested in the issues as intrigued by the language and rhetoric of the times...It wasn't like another world but the same one only with more urgency, and the issue of slavery wasn't the only concern...You wonder how people so united by geography and religious ideals could become such bitter enemies. After a while you become aware of nothing but a culture of feeling...It's all one funeral song...

Just like other art forms, fiction may be – and often is – the product of substantial investigation. It may frequently use protagonists that have left their marks on what, for want of a better term, we think of as 'real' history. Take, for instance, Mario Vargas Llosa's minute novelistic excavation of dictatorship in *The Feast of the Goat* (2001). This work calls on a tradition of Latin American writing on *caudillismo* exemplified by earlier novels such as Alejo Carpentier's brilliant *Reasons of State* (1997). Vargas Llosa has written a novel, to be sure. But its reconstruction of Rafael Leonidas Trujillo's corrupt inner circle in the Dominican Republic, the dictator's pernicious effects on a whole society, and his lasting legacy for some of his victims and erstwhile collaborators, is plainly the product of the library and of discussions with informants.

Sometimes, the debt commonly owed to research is made explicit, Dylan style, when a fiction writer pays tribute to the sources used. At the end of his novel on Irish republicanism and the Easter Rising of 1916, *A Star Called Henry* (Doyle, 1999), and in its US-based sequel centred on bootlegging, the Jazz Age and the Depression, *Oh, Play That Thing* (Doyle, 2004), Roddy Doyle provides his reading lists. As the books have historical protagonists with whom Doyle's invented characters interact, we are being told – in effect – that some of what we read is indeed believable because it can be accessed extra-textually. But we do not therefore imagine that what we have read is something other than fiction.

In *The Glass Palace*, Amitav Ghosh's post-colonial epic, set in India, Malaysia, Myanmar and Thailand, the author provides an unusually revealing example of how a fiction-writer may account for the sources used. Ghosh was once an anthropologist for whom the professional convention of providing a methodological appendix would be second nature. Instead, as a novelist, he gives the equivalent – four pages of 'author's notes' (Ghosh, 2001, pp. 549–52). These detail the five years of research behind the book, which involved not only voluminous reading but also extensive travel and interviewing. The autobiographical trigger for the whole venture, we are told, was the author's father's own memories. A key protagonist in Ghosh's tale is Dinu, a photographer for whom photography involves a kind of mystical communion both with the people and things whose images he takes. The practice of photography – so central to the ethnographic enterprise – is minutely described in the novel. The descriptions are underpinned by information from an expert informant, cited in the author's notes.

Despite the conventions that demarcate it as fiction, therefore, the novel often has a complex relation to a world of narrative reconstruction which may rest on the deposits provided by eyewitness testimony and the evidential support offered by verifiable documentation.

III

The anthropologist James Clifford notes that, in its pioneering days, fieldwork had to distinguish itself from literary and journalistic accounts of travel, as a particular practice, and as producing its own kind of knowledge. The separation of genres – and thereby the creation of an intellectual status and a sacral frontier – was part of the production of a professional, anthropological identity. However, Clifford (1997, p. 66) believes that now the old fences are being dismantled. He remarks: 'the current "experimentalism" of ethnographic writing is a...renegotiation of the boundary, agonistically defined in the late nineteenth century, with "travel writing"'. Presently, he argues, elements of the travel narrative are returning to anthropology proper. He cites in evidence the descriptions of 'the researchers' routes into and through "the field"; time in the capital city, registering the surrounding national/transnational context; technologies of transport (getting there as well as being there); interactions with named, idiosyncratic individuals, rather than anonymous, representative informants' (Clifford, 2001, p. 67). It is, at least in part, a shift from an impersonal, scientistic mode of expression to the rediscovery of the first person.[2] Just how far the anthropological ego should intrude in writing ethnography has been a matter of extensive controversy in the profession (Macdonald, 2001).

Nonetheless, Clifford's description of the self-aware, travelling anthropologist characterizes much of W.G. Sebald's work, in which the reporter-traveller, who is sometimes a note-taking researcher, follows his itinerary. Routine reference to modes of transport, the reproduction of photographs, detailed accounts of conversations, and minute descriptions of locales are central to his style. '*Literary* ethnography' of this kind is anchored in reporting encounters, situations and objects observed, in a style akin to anthropology – and, not surprisingly, to the travelogue.

I intend, by using the term 'literary ethnography', to emphasize that we are confronting the excitement and uncertainty of a borderland – or more strictly, because there is no single, indisputable, line to be drawn, of border*lands*. To think of works as interstitial is to resist imposing a generic categorization. To impose a generic choice is to engage in ordering, in making things clear and distinct. But the very act of needing to categorize, of discussing the appropriateness of one label as against another, suggests that the material is inherently refractory. At the margins, how we plump for one or other description of a literary endeavour is ultimately arbitrary (though it may be justifiable on the basis of this or that criterion). It is the compelling nature of such marginal choices that contributes to the intrigue that surrounds a work that is hard to classify. Such works are disruptive and disturbing. They make us think about how we should read them and respond to what they are saying.

In reflections on the French writer Georges Perec, the distinguished sociologist-ethnographer Howard Becker has argued that some of Perec's work is 'proto-ethnographic'. While not sociology *per se*, it is implied, such writing shares common ground and techniques with the researcher in the field, not least in its observational qualities and minute description.

Perec's title to proto-ethnography was soundly established in his own working life. David Bellos' (1995) magnificent biography shows how Perec was grounded in market research, with extensive experience of tape-recording and transcribing interviews. He was, in the parlance of the day, an amateur *pyschosociologue,* well versed in the Marxist sociology of everyday life, although without formal training in social research. He also worked for many years in scientific documentation and information retrieval for a research laboratory, while engaged in his literary pursuits, and was deeply familiar with the production of academic work (the referencing conventions of which he satirized to great effect).

In one autobiographical work, *W or the Memory of Childhood*, Perec addressed his own displacement from Paris to the countryside, as a Jewish child under the Nazi occupation. His identity was concealed at his Catholic school; he was baptized and his surname became 'Breton'. He had to forget that he knew Yiddish and Hebrew. Perec's father was killed in 1940 while serving in the French army and his mother was deported to a concentration camp, never to be seen again. Speaking of *W*, Perec said that he had written it 'because it is a way of bearing witness' and that the concentration-camp system had placed a terrible burden 'on huge numbers of people during the war – and long after' (cited in Bellos, 1995, p. 564).

Unquestionably, Perec is a key precursor of Sebald, in whose *oeuvre*, loss of identity, language and parents, and the long-term psychic disturbance that ensues, are likewise central themes, particularly so in the last novel, *Austerlitz*. Perec was an internal exile who survived the occupation and deportations; however, his parents did not. Although Perec's book is autobiographical, Bellos (1995, ch. 52) has shown in considerable detail how he deliberately throws false trails and renders details inaccurate. Bellos reads Perec as both telling his story of loss, and as unwilling and unable to disclose everything he had to say about it.[3]

A famous passage (with clear echoes in Sebald's work) concerns the departure of the young 'Jojo' Perec from Paris to relative safety in the Vercors, under Vichy control. This is the last occasion on which he sees his mother:

> My mother accompanied me to the Gare de Lyon. I was six years old. She entrusted me to a Red Cross train that was leaving for Grenoble, in the unoccupied zone. She bought me a comic, a Charlie Chaplin, on the cover of which you saw Charlie, his cane, his hat, his shoes, his little moustache, parachuting. The parachute was attached to Charlie by his braces. (Perec, 1975, p. 80; author's translation)

Bellos (1995, p. 547) calls this an 'error-trap' and 'an almost certain fabrication' because Chaplin's image was banned in Nazi-occupied Europe. While autobiography is invariably selective in what it discloses, in Perec's case, it also plays games by wilfully providing details that falsify and can be shown to be false. This destabilizes the idea that autobiography should try to validate what it tells. Does introducing false trails make it a kind of fiction?

Viewed in this light, three of Sebald's works *Vertigo*, *The Emigrants* and *The Rings of Saturn*, raise problems of classification. A central issue for literary critics has been the extent to which they are autobiographical. Or, putting it differently, we are asked to address the tensions between the fictional and the non-fictional. These points apply no less to a fourth work, his major novel *Austerlitz*. While plainly a work of literary fiction, this actually *presents* itself as a quasi-ethnographic travelogue, as a self-styled 'report', and we cannot doubt that the author has indeed been a traveller-researcher who features his own itinerary in his fiction. While we know that he abhorred confessional writing on the grounds of a self-protective privacy, we also know that he constructed a narrator's persona with more than a passing connection to his biographical self. Occasionally, Sebald would reflect directly on the motivations behind his work. For him, making a fundamental reckoning with World War II was an axiomatic starting point for coming to terms with the past:

> At the end of the war I was just one year old, so I can hardly have any impressions of that period of destruction based on personal experience. Yet to this day, when I see photographs or documentary films dating from the war I feel as if I were its child, so to speak, as if those horrors I did not experience had cast a shadow over me, and one from which I shall never entirely emerge. (Sebald, 2003, pp. 70–1)

This points clearly to the crucial significance of photography (and film) for Sebald's writing: mediated images – for him, as for all of us – are a key means of accessing the past and constituting 'memory' and hence an orientation to the past. We shall return to this point.

IV

Sebald is a quintessentially European writer. Not only is his sensibility rooted in – particularly – central and western European literary traditions, it is the European continent that figures most in his work. Even when he writes about the United States or Africa or China, it is mainly through the prism of assessing the European impact on these areas through colonization or migration. The sense of Europe often evoked is of a continent whose borders shade into one another. Europe is a cultural area, a space of common heritage – although one in which much blood has been spilled because of national and ethnic conflicts. A consequence is that crossing borders can at times be a problem, especially so for those trying to escape persecution or death.

All of Sebald's literary works are representations of movement through space and time to which the recounting authorial presence is central. To travel in the contemporary world involves crossing frontiers where a question mark always hangs over your successful entry and exit. The political philosopher, Seyla Benhabib (2002, p. 86), remarks that how the borders of a polity are regulated are a means to our understanding of the mentalities that define 'us' and 'them'. 'Sovereignty entails the right of a people to control its borders as well as to define the procedures for admitting "aliens" into its territory and society.'

This controlling obsession has lost nothing in the era of 'globalization', it would seem. It remains at least as vital today as it was during that of the World Wars and the Cold War, during what Eric Hobsbawm (1995) has called the 'short twentieth century'. Indeed, the upsurge of migration that has accompanied globalizing trends has ensured that the watch over Fortress Europe has become ever more vigilant. The management of borders remains the subject of continuing state control and surveillance, especially following the dissolution of the Iron Curtain, with illegal immigration, people- and drug-trafficking high on political agendas inside the European Union (EU). Times of high tension about security threats due to terrorism have had far-reaching consequences for how states combine to police and gather intelligence (Statewatch, 2003a; 2003c; 2004).

The social anthropologist Orvar Löfgren (2002, p. 251) writes that understanding national frontiers is part of the 'pedagogy of space' to which we have been subjected in the modern era. Legitimate border-crossings today require the right kind of documentation that classifies you in terms of your entitlements and rights. To be stateless is a kind of disgrace with the most profound consequences for your freedom of movement, and not least, for your identity. Of course, the world of the passport as we know it is only about one century old and more than anything a product of the need to control movement during World War I (Löfgren, 2002, pp. 254–5). Increasingly, surveillance is being linked to information technology, the latest manifestation being the encryption of biometric data and its adoption for border control in the USA (Statewatch 2003b). In general, through their exercise of controls over transit, ingress and egress, states exert a conditional, and conditioning, power over their inhabitants' fates, creating obstacles for those who try to slip through the doors of the realm clandestinely.

In *Vertigo*, Sebald throws his own credentials as an authorized traveller into question. The book is partly about the author's quest for his own identity. It is also, in part, an ironic contemporary evocation of the grand tour. So when Sebald tells how his passport is lost at a small hotel on Lake Garda in Italy in 1987, this is a particularly pregnant and revelatory moment. For several pages the author gives a droll account of the drama that ensues in the hotel as the owners seek the lost passport, accompanying the hunt with much hand-waving. Eventually, they realize it was given to 'another German'. Sebald is despatched to the local police station, accompanied by the hotel owner's wife, to obtain some travel papers:

> The *brigadiere*, who wore an immense Rolex watch on his left wrist and a heavy gold bracelet on the right, listened to our tale, sat down at a huge old-fashioned typewriter with a carriage practically a metre across, put in a sheet of paper and, half murmuring and half singing the text as he typed, dashed off the following document, which he tore out of the rollers with a flourish the moment he had completed it and read it over once more for good measure, handing it first to me, who was rendered speechless by this virtuoso performance, and then to Luciana for signature before endorsing it himself first with a circular and then with a rectangular stamp. When I asked the *brigadiere* if he were certain that the document he had drawn up would enable me to cross the border he replied, faintly irritated by the doubt implied in my question: *Non siamo in Russia, signore*. (Sebald, 2000, pp. 101–2)

This vivid, detailed description is accompanied by a photograph of the document, which is evidently authentic. The story conveys 'Italianicity', as Roland Barthes (1977) would have said. It is manifest in the description of the *brigadiere*'s theatricality; while a caricature, this signifies the importance of state endorsement for travel and indeed the anxiety of the traveller to receive it. And in the dismissive reference to Russian despotism, the vignette calls into existence the well-worn distinction between the free and the communist worlds. Sebald grew up in a divided Germany. However, the Cold War and its fracturing impact on Europe are largely absent from his evocation of the continent. The above quotation notes a rare, merely allusive, intrusion of the Iron Curtain, which is only ever at the margins of Sebald's concerns (Schlesinger, 2004b).

The lack of a passport continues to dog Sebald as he makes his way to the German consulate in Milan to regularize his position. In a 'lite' version of illegal itinerancy, he is treated with intense suspicion at another hotel when he cannot produce his papers (Sebald, 2000, p. 110). Sebald then describes, again with a touch of Felliniesque pastiche, what happens in the consulate:

> Although it took a long while until my identity had been established, by several phone calls to the relevant authorities in Germany and London, the time passed lightly...At length, a short, not to say dwarfish consular official settled himself on a sort of bar-stool behind an enormous typing machine in order to enter in dotted letters the details I had given concerning my person into a new passport. Emerging from the consulate building with this newly issued proof of my freedom to come and go as I pleased in my pocket, I decided to take a stroll around the streets of Milan for an hour or so before travelling on...(Sebald, 2000, p. 115)

Official authentication to move as a contemporary *flâneur* through political space, then, is the necessary precondition for actually crossing frontiers. However, as states issue passports – and withdraw them – it is a conditional freedom. Even having the right credentials does not always mean that the passage is without incident. Nor does the possession of citizenship automatically abolish the sense, however vague, that crossing the frontier may often prove to be a significant, anxiety-producing act. This may apply, as Löfgren (2002, p. 253) points out, even to the transit from Sweden to Denmark, which for him as a Swede 'has lost much of its drama'. Being a passport-bearing citizen does not always make you feel easy, as he goes on to observe:

> The fascination or strength of border crossings, which still exists in a world of deterritorialization and deregulation, has to do with the fact that in a world where fewer and fewer identities are based on the clear-cut pedagogy of space, the nation-state still tries to provide an absolute space: Sweden or the United States starts here!...At the border, the selective nationalizing gaze is scanning the terrain for alien elements, fluids, objects, individuals, influences. What must the nation be protected from - in a given situation and at a given time in history? (Löfgren, 2002, p. 271)

In *The Emigrants*, Sebald relates his own encounter with the 'nationalizing gaze' when he first arrived in England in 1966, landing at Manchester:

> Although only a scant dozen passengers had disembarked at Ringway airport from our Zurich flight, it took almost an hour until our luggage emerged from the depths, and another hour until I had cleared customs: the officers, understandably bored at that time of the night, suddenly mustered an alarming degree of exactitude as they dealt with me, a rare case, in those days, of a student who planned to settle in Manchester to pursue research, bringing with him a variety of letters and papers of identification and recommendation. (1997, p. 50)

Again, Sebald notes his need for credentials. But his minor difficulty is a deeply ironic counterpoint to stories of other migrants that he recounts. *The Emigrants* is a collection of tales that shows how even minimal impediments when crossing frontiers may vitally affect the well-being - not to speak of the survival - of its characters. Sebald writes of men - such as his great-uncle Adelwarth - who sought their fortunes abroad (what the state and media now call 'economic migrants'). Even more poignantly, he tells of those who have escaped genocide or racial persecution (in the current, received parlance, they are 'asylum-seekers'; once they were 'refugees'). For characters such as the artist Ferber (who escaped Nazi Germany when sent abroad as a Jewish child, by parents who subsequently perished), the need to cross the border was quite simply a matter of life or death. Acquiring British citizenship likewise transformed the fate and fortunes of Dr Selwyn, fleeing the Russian pogrom.

V

It is in his novel, *Austerlitz*, that Sebald arguably perfects his literary ethnographic project. This 'report' (as it is sub-titled in the original German edition) embodies the considerable documentary research evident in all of the other books, as well as the author's compulsive information-gathering itinerancy on the Continent. Although it grounds the story in part in verifiable fact, and supports the narrative with the same kind of ambiguous photographic 'evidence' familiar to Sebald's readers - of which, more later - unlike all the other works, this is intended to be read wholly as a work of fiction. And yet, because the author saw it as 'an alternative holocaust museum' (Krauthausen, 2001, p. 2), it is also a political critique of the Nazi genocide of Europe's Jews.

Sebald had earlier written about characters bearing names that obscured, rather than revealed, their true – possibly composite – identities. The eponymous protagonist of *Austerlitz*, Jacques Austerlitz, is likewise partly inspired by a true story (Jaggi, 2001, p. 3). The account derives an atmosphere of documentary authenticity from its evident grounding in the author's own experiences. We do not doubt that – as part of his preparation for writing this novel – he has visited the locations of which both the narrator and Austerlitz speak; moreover, some of the life-events that are recounted by the narrator help to build the book's relation of intimacy with the reader. There is one footnote of apparent personal reminiscence, a further factualizing device (Sebald, 2001, p. 8). Sebald also cites several books as either part of Austerlitz's, or the narrator's, reading.

Echoes of Sebald's other exile stories are perceptible too. If we are inclined to believe the literary truth of those accounts, we are also so inclined in this case. *Austerlitz* is particularly clear in how it characterizes the questions of memory and identity. Its eponymous protagonist is an academic whose articulation of his situation – and his procedures for establishing the truth about himself – are familiar to those of a university culture. In his quest for the truth, Austerlitz makes use of academic research methods: reading, interviewing, fieldwork, library work, archival searches. These practices are described in detail, as befits any reflexive account of one's methodology. It is conventional for fiction-writers to separate research from the creative imagination, and to place references to the investigative groundwork in an 'author's note'. However, in Sebald's literary ethnographic mode, the investigation becomes an integral part of the text, not least because that is part of the *style* and the intended effect.

It is in *Austerlitz* that Sebald most fully explores the political nature of movement through space – and its impediments. Over a period of some thirty years, the narrator moves fluidly around western and central Europe in the course of writing his 'report'. In acute contrast to his own freedom of movement, he writes of people compelled to uproot themselves by the Nazi war-machine. Or of those forced to flee because they are trying to escape the Nazis' clutches. Or again, of others whose countries have been occupied, thereby forcing them to stay put. Austerlitz, who recovers his own lost identity only in middle age (he is a Czech Jew, sent at the age of five to England on a *Kindertransport*) is the embodiment of flight, of the exile or refugee.[4] His mother, he discovers late in life, was deported from Prague to the concentration camp at Theresienstadt, from whence every trace was lost. His father escaped to France, probably handed over to the Nazis by the Vichy regime, and may have been exterminated in Lithuania. Their fates are uncovered only through the survival of his parents' friend in Prague, Vera, who is able to provide relevant testimony. Sebald's work underlines how crucial it is to have the right passport at a time of crisis, persecution and repression.

VI

A characteristic feature of many anthropological accounts – as indeed of biographies, histories and travelogues – is to include photographs of key scenes and persons referred to in the narrative. Text and image are meant to illuminate, and mutually modify, each other. It is no accident that Sebald thought of *Austerlitz* as an 'alternative holocaust museum'. The metaphor directs us to think about how, in the museum, the photograph, along with its captions and the other texts that describe objects on display, are means to 'construct history', in James Clifford's phrase. They typify and recontextualize cultures and their products for the visitor. The notes that we find in museums, moreover, 'are not, properly speaking, descriptions of the objects to which they refer. Rather they are interpretations that serve to open a meaningful space between the object's maker, its exhibitor, and its viewer, with the last-named given the task of intentionally, actively, building cultural translations and critical meanings' (Clifford, 1997, p. 136). This process applies no less to how photographic reproductions and their captions are intended to work on the spectator.

It is worth underlining this point. Although Sebald conforms to the ethnographic norm by displaying photographs and other images in his books, he also departs from it as he completely eschews the use of captions. By doing this, he wilfully removes cues to interpretation normally provided for the reader. We are therefore required to do more work. We are also invited to be puzzled about the intentions behind this textual strategy. The use of photography signals the ethnographic dimension of his texts while, at the same time, the lack of conventional signposting denies it. Refusing to caption in line with the usual practice and expectations of the social sciences emphasizes that this is a particular *literary* ethnographic choice.

Susan Sontag (1979, p. 8) has observed that '[t]hrough photographs, each family constructs a portrait-chronicle of itself – a portable kit of images that bears witness to its connectedness. It hardly matters what activities are photographed as long as photographs get taken and are cherished.' To carry photographs of one's nearest and dearest is normal for the migrant. For exiles ignorant of their origins, such commonplace photography is a lost treasure and therefore – once recovered – has the most extraordinary evocative power. Sebald invokes this with particular force as he recounts how Austerlitz's gradual retrieval of his identity brings to light the sole surviving image of himself as a youngster. This – a portrait of the child-cavalier on his way to a fancy-dress ball in Prague – is the iconic photograph of *Austerlitz*. It is displayed on the book's front cover, as well as at the appropriate narrative point in the text. By a happy accident, the photograph is found by Vera, the old family friend. This is how Sebald describes its hand-over, in a lengthy passage, from which the following lines are extracted:

> Minutes went by, said Austerlitz...before I heard Vera again, speaking of the mysterious quality to such photographs when they surface from oblivion. One has the impression, she said, of something stirring in them, as if one caught small sighs of despair...as if the pictures had a memory of their own and remembered us, remembered the roles that we, the survivors, and those no longer among us had played in our former lives. Yes, and the small boy in the other photograph, said Vera after a while, this is you, Jacquot, in February 1939, about six months before you left Prague...The picture lay before me, said Austerlitz, but I dared not touch it...I did recognize the unusual hairline running at a slant over the forehead, but otherwise all memory was extinguished in me by an overwhelming sense of the long years that have passed. I have studied the photograph many times since...(Sebald, 2001, pp. 258–9)

Uncaptioned, the photograph's initial interpretation is provided by Austerlitz, with Sebald constructing the narration both to underline the 'mystery' of what it tells and the autonomous life that photos seem to have. But this does not exhaust the process of wondering what it might mean. We, the readers, ask of *whom* is this a photograph, and where and when was it taken? How did Sebald acquire it? And why did he choose to tell the story in precisely this way – selecting precisely *that* image?

Removed from its particular context, and functioning more as an icon than as a purely referential image, the child-cavalier comes to stand for *all* those children who have lost their parents and their histories. The photograph also stands for the cards of identity that such exiles must seek to reassemble in order to remember who they once were. Indeed, such an image, once retrieved, may be less part of remembering who one is than an indispensable means of *discovering* it. The photograph is an entry point for those trying to identify who they have been in another life, a means to reappropriate the web of associations and relations in which they once lived.

For Sebald, the emotional charge carried by photography also has an ethical dimension. In *The Emigrants*, the key precursor to *Austerlitz,* the narrator concludes the book by telling of a photographic exhibition he once viewed in Frankfurt. This was a record of the Litzmannstadt ghetto taken by an Austrian official who had worked there. It provides a scrupulously detailed record of a captured tribe that is destined to disappear. One of the photographs, we are told, was particularly arresting. It was of three women weaving in a ghetto workshop:

> Who the young women are I do not know. The light falls on them from the window in the background, so that I cannot make out their eyes clearly, but I sense that all three of them are looking across at me, since I am standing on the very spot where Genewein the accountant stood with his camera. The young woman in the middle is blonde and has the air of a bride about her. The weaver to her left has inclined her head a little to one side, whilst the woman on the right is looking at me with so steady and relentless a gaze that I cannot meet it for long. (Sebald, 1997, p. 237)

That gaze, we understand, is one of accusation, perhaps even of defiance towards the objectifying lens. Now, extruded from the secrecy of the photographer's closet and fully out in the public domain, the photograph becomes mute testimony to suffering. The narrator's response (contrary to that of the photographer) is to experience guilt – and, as we see from the development of Sebald's *oeuvre*, this means taking on responsibility for telling stories of victimhood.

VII

Whereas Sebald uses his literary construction of exile to focus intensely on the individual case, there are other options, not least the representation of *collective* experience. This is the approach taken to Jewish exile in the major photo essay by Frédéric Brenner (2004a; 2004b). His two volume study, *Diaspora: Homelands in Exile,* combines a fine-grained interrogation of specific cases with a far-flung diversity of location. The underlying conceptual

connection between 'diaspora' and 'exile', therefore, is this: due to dispersal, the exilic condition may take many forms and conditions. The exilic condition is *inherent* in diasporic life.[5] Return to the Land of Israel, moreover, does not necessarily terminate the exile, as in theory it should. For many, exile simply continues in another place because the returnee still comes from somewhere else, and that somewhere has shaped his or her identity and difference. For Brenner, as for Sebald, exile seems inescapable. But while Sebald's vision is tragic and melancholic, Brenner's is hopeful.

Brenner, with a background in social anthropology, film-making and writing, as well as photography, conceived his diaspora project as a visual anthology that gathers and assembles 'the multiple fragments of exile', producing a pictorial account before the final disappearance of what has been captured. Brenner's information-gathering travel spans 1978–2002. As with Sebald, the author is the recorder (and also the arranger, sometimes the impressario) of what he gathers. Whereas Sebald's itinerary is followed mostly in western and central Europe, Brenner (2004a, p. ix) has travelled the world in 'an odyssey that would lead me over a period of twenty-five years from Sarajevo to Calcutta, from Rome to New York'. Sebald's pre-eminent concern is with the impact of World War II on the Jewish exile and also on the German nation. Brenner is certainly concerned with the Holocaust but also with diasporas whose trajectories date at least from the Spanish and Portuguese Inquisitions of the fifteenth and sixteenth centuries (Sachar, 1994). What he found on his 'odyssey' raised far-reaching questions about diasporic identity (and perhaps, about how we construct our very notions of collective identity):

> When I'd started out, my project had been, unavoidably almost ethnographic in nature, if not in intent. But the more I progressed, the more I was forced to abandon the myth of 'One People'. I was searching for what I believed in: continuity. I found only discontinuity. And the more Jews I met, the less I understood what a Jew looked like. (Brenner, 2004a, p. ix)

Brenner's project, therefore, rejects an essentialist conception of identity. There was no single tribe to be found. As Stanley Cavell (2004, p. 133) comments, Brenner's work provides 'a guide in how to escape making fixed, metaphysical identifications of the possibilities of others'. With two volumes each around one square foot in size, Brenner's output hardly resembles Sebald's. At a glance, his books would seem destined for the coffee table or the outsize bookshelf. And yet, there is much common ground with the project of Sebaldian literary ethnography.

The first volume is subtitled 'Photographs', and apart from a short introductory essay by the author, contains a sequence of black and white images, each accompanied by 'brief descriptions, pointing to an identity, a date, a location' to which you bring your own apparatus of reading (Dayan, 2004a, p. 34).[6] The photographs do not stand in isolation, as a selection is cross-referenced to the second volume, 'Voices'. At less than half the length, this reproduces a selection of photographs from the first volume. Each of the photographs is accompanied by several written texts, mostly commentaries, though some are interviews and others testimonies.

To produce these, Brenner assembled a cast of 26 contributors, most also Jewish, but from diverse backgrounds and intellectual standpoints.[7] The construction of the diasporic object, therefore, is analysed and debated by a group of (mainly) diasporic subjects, engaged to read not only through their own disciplinary frameworks but also through their own exilic experiences and memories.

In his introduction to 'Voices', Tsvi Blanchard (2004b, p. vii) describes the work as a 'photographic Babylonian Talmud'. The captions 'leave room for exegesis and imaginative association'. Following the ancient Talmudic principle, Brenner offers a text for commentary and has organized a range of authorities to undertake this task. The reader is invited to join in, and by his or her own act of interpreting, to adjudicate between others' interpretations and so to take part in a grand hermeneutic exercise. The interpretations offered are, variously, poetic, personal, aesthetic, historical and sociological. As readers, we can make a choice: to disregard the 'voices' and be our own interpreters; or to hear others' voices and to engage in communal interpretation. Brenner's project accords closely with the idea of a 'Jewish ethnography', conceived not necessarily as an ethnography of Jews but as potentially universal in application. Daniel Boyarin (1992, pp. 74–75) has developed this idea. He suggests that the 'dialogic play of interpretations within the bounds of an implicit ethical framework' offers a 'Jewish interpretive model of multiple and diffuse authority, of dialogue with and *through* a narrative, textual source whose potential meanings are never exhausted, and hence never fixed'.

Any publicly articulated interpretation of photographs, however cryptic, involves a telling by the interpreter. The photographic essay conceived Talmudically thus becomes the stimulus for incessant narrative. To view is to pick up one tale and produce another. In that respect, interpreting the images is consistent with a conception that links the production of exilic photography to the stories that prompted it in the first place, much as Sebald's work is rooted in emigrants' accounts and Perec's in his own experience (though both went in for careful obfuscation). Brenner explains that:

> At the origins of every photograph in this book, there is a story. Not an image, but an anecdote, a name discovered or a place described, which I heard and later recounted, and which recounts a part of me and helped me to reclaim the fragments of my own story. (Brenner, 2004a, p. x)

The photographer therefore consciously inserts himself into the history of the Jewish people and invokes his own exilic ancestry. By photographing diasporas he is taking on the sacred task of conserving and transmitting memories that define an identity – while at the same time questioning its claims to coherence. It is a strategy of reporting *from within* and exploring one's own sense of oneself by confrontation with an assumed sameness that proves – in the end – to be a difference. Sebald's ethic of responsibility, by contrast, is quite distinct. Leaving Germany as a young man, he entered into a self-imposed exile and it was this condition that enabled him to travel *from outside* to become, in the end, a kind of Jewish writer.

VIII

Brenner's photographic subjects are located across five continents, range from young to old, have diverse sexual orientations, different religious practices or none, include rabbis (of both sexes), aristocrats, the rich, professionals, the politically engaged, business people, merchants, small traders, poor industrial workers, craftsmen and tribespeople - and more. They are photographed in their homes, at work, at prayer, in a range of urban and rural settings. The question of how such diverse people can be connected - and *think* themselves to be so - is therefore posed on every page.

Given the space at my disposal, I have selected just one sequence of photographs and commentaries for a brief discussion; it is close to Perec's and Sebald's theme of displacement and the Holocaust. My example is a classic instance of ethnographic practice, as it deals with an enigmatic, carnivalesque ritual (the very stuff of anthropology). Moreover, it is hard to decipher.

In his introduction to *Diasporas*, Brenner writes of the village of Tykocin in northern Poland, which he visited because of a story he was told. In August 1941, shortly after the Nazi invasion, the Jewish villagers were taken to the nearby forest (with the active collusion of Polish neighbours) and killed. The complete disappearance of Jews left a void and eventually a strange practice was invented to fill it. This is a re-enactment of the Jewish festival of *Purim*, which celebrates Jewish deliverance from a genocidal plot, hatched by Haman, a close advisor of King Ahasuerus. The plot was foiled through the benign intercession of Queen Esther, the king's favourite wife, and Haman was himself destroyed.

Brenner gives this case unique attention: the story of Esther is printed, without comment, immediately before his introduction to *Diasporas*. He introduces the topic as follows:

> February 2002. Tykocin is a town without Jews, where the inhabitants, from the priest to the truck driver, the teacher and schoolchild, the baker and mechanic, dress up as the Jews whom they never saw but whose memory they wish to preserve. This is Purim - the only festival in the Jewish calendar that is as much about inversion as it is about remembrance. The Catholic Poles put on a skit or *Purimspiel*. The next day, they march in procession through the streets of Tykocin to the synagogue, where they hang an effigy of Haman. Here we now have Purim, but a Purim without Jews. The story of Purim in the scroll of Esther prefigured the Shoah, the intent 'to destroy, to kill, to cause to perish all Jews, both young and old, little children and women, in one day'. The Poles of Tykocin helped it come to pass. Today, therefore, the children and grandchildren of those Poles celebrate Purim twice-reversed. (Brenner, 2004a, p. xiv)

Those who perform the 'parody' (often for Jewish visitors from abroad) know nothing of the massacre, or, if they do, are silent about it. Brenner (2004a, p. xv) admits that he does not quite know how to make sense of this '[a]mnesia dressed up in the trappings of memory', or what it implies for the identities either of actors or spectators.

We are presented with eight photographs. A characteristically tall, tightly cropped sequence of seven consists of head and torso portraits. Six are of people, three in their everyday clothes, three dressed up as performers. Nobody smiles. The seventh portrait, which divides the others into two groups of three, is, in Daniel Dayan's words (2004b, p. 123), of 'a totem of sorts', an 'African mask of a Polish Jew' – we are, indeed, right in the territory of folkloric cultural production.

The eighth photograph is a wide shot of the *Purimspiel* on the village green, showing the hanging of the wicked Haman, starkly silhouetted under a pale, late winter's sky. The other players – including the king and queen, and the queen's uncle, Mordechai – stand near the gibbet, slightly windblown, either smiling or making victory gestures at the death of the would-be murderer. They are in cobbled-together fancy dress.

Returning to the first three portraits, the first is of a middle-aged Catholic woman, a convert to Judaism, whereas the third is of a young, former Catholic who discovered his Jewish origins, and entered a Jewish seminary. Both plainly deal with identity switches, raising questions of how the Jewish and Polish dimensions of identity interrelate, and may change. The third, more enigmatic, is of a bearded man connected to Solidarnosc, son of communist ideologues. Is he Jewish? Does he *think* he is? Placed between the other two, and divided from the three performers by the 'totem', we assume that this is in some sense a group with a 'Jewish' identity, whether embraced, inherited or denied.

The other three portraits are of the female director of the Tykocin Jewish museum, the male star (playing Mordechai) of the *Purimspiel*, and a Catholic priest. All are cast as 'Jews'. Dayan sees the museum director as a kind of 'scarecrow' got up as a Jewish ghost; the priest, uncomfortable with his side-curls as a 'cross-dresser caught off-guard'. The third character also looks inauthentic. His small, fringed, prayer undergarment is on uncustomary display, and a patently false beard sits under the chin. These are approximations to, rather than appropriations of, character.

Brenner has offered no firm interpretation himself. Here, I will consider two of the texts printed in 'Voices' to accompany the images. One is an excerpt from a letter by a rabbi and Judaic scholar, Yehiel Poupko, who inspired Brenner to go to Poland. Poupko (2004, p. 122) argues that the sudden mass disappearance of Jews in the Holocaust left a negative presence in the places they once lived and that this demands a 'photography of absence'. The *Purimspiel* plays out precisely the disappearance of which he writes. The photography of absence, therefore, is able to document a perverse presence. For his part, Daniel Dayan (2004b, pp. 122–3) interprets the performance – in which 'Jewishness is a visual style' – as a kind of 'Passion Play', thereby annexing it to the Christian tradition of dramatizing Christ's crucifixion. He puzzles over the meaning of the ritual and asks what drives people, some of whose forbears were responsible for destroying another people, to take on their guise and perpetuate their name.

From the perspective of this essay, the *Purimspiel* is about three things. First, the consequence of the *failure* to go into exile – namely, death; second, how the disappeared are apparently reincarnated through a perverse working out of memory; and third, how the process whereby this seeming restoration occurs becomes itself a fitting object of ethnographic study.[8]

IX

This essay is merely a preamble to a much larger study. I have suggested that there are powerful connections between the condition of exile and the ethical obligation to document what this means. The recourse to different forms of ethnography provides a means of addressing this compulsion. As we have seen, the ethnographic eye can variously shape autobiography, reportage, genre-bending fiction and photography. That does not exhaust its representational potential, by any means. At the heart of the matter are questions of how we constitute memory and construct identity, both as individuals and as members of collectivities. Exilic ethnography, generously understood, is surely central to this, often fragile, process.

Acknowledgements

This chapter was originally published in Social Science Information/Information sur les sciences sociales, Volume 45, No.1, March 2006, pp. 53–71. I am grateful to the editor-in-chief, Anne Rocha Perazzo, for permission to republish. My sincere thanks also go to the Maison des Sciences de l'Homme in Paris, where I conducted some research on this project in 2002 and 2005. I am further grateful to Sharon Macdonald and Martin Peterson for their detailed comments. My appreciation, too, to my colleagues on the ESF's 'Changing Media, Changing Europe' Programme (2000–2004) as well as to all those others who have responded so encouragingly to this essay.

Notes

1. In Schlesinger (2003); see also Schlesinger (2004a; 2004b).
2. In other – popular – cultural arenas we would link this to the confessional turn now inseparable from the dubious status of celebrity, or the come-clean interview and kiss-and-tell revelations in sexual and financial scandals, or the transformation of obscure individuals into temporary household names by 'reality television'.
3. The 'W' is interpreted by Bellos as an inverted 'M', standing for the lost *mère*, or mother. Perec was an accomplished linguist and played language games. I believe the title has yet another meaning. The 'W', in French, is pronounced 'Double-Ve'. In German, the 'W' is pronounced 'Ve'. The noun *das Weh* (pronounced 'Ve') is sorrow or grief. Surely Perec is alluding to his '*Double-Weh*', his unmitigated sorrow at being an orphan?
4. Sebald's indebtedness to Perec cannot be doubted when he narrates Austerlitz's recollection of his departure from Prague as a child. Austerlitz tries to recover mental images of his mother but fails – but he does remember that she bought him a Charlie Chaplin comic for the journey (Sebald, 2001, p. 308). In *Vertigo*, 'W' is the birthplace to which Sebald returns.
5. My thanks to Daniel Dayan for drawing Brenner's work to my attention during one of our ambulatory Oslo conversations. Dayan (2004b, p. 123) has briefly noted the connections between Sebald's and Brenner's projects.
6. There is an exception. On pages 285–7 are the only two colour pictures in the book. Commissioned by the Vienna State Opera, these are of fabric, incorporating as a motif the yellow star that Jews had

to wear as identification under the Nazis. This material was hung to chilling effect as a curtain in the Opera.

7. Volume 1 has 262 duotone photographs, whereas Volume 2 reproduces 60 of these, in the words of the blurb, 'along with commentaries by such leading contemporary thinkers such as André Aciman, Jacques Derrida, Carlos Fuentes, Barbara Kirschenblatt-Gimblett, Julius Lester, George Steiner, Avivah Gottlieb Zornberg'.
8. These last reflections have produced *another* layer of interpretations, in the Talmudic spirit, sent to me as private communications in February 2005. The Swedish Europeanist, Martin Peterson, writes that the Tykocin case 'has its equivalents in the Jewish theatre in Warsaw where performances with non-Jewish actors are in Yiddish and more often than not are dealing with heart-breaking stories from typical pogroms...The element of magic realism is palpably present. Similarly, the new Jewish Museum in Berlin...also performs theatre plays by Singer et al. played by both Jewish and non-Jewish actors. These performances have nothing to do with any maintaining of tradition but rather with a creation anew of a fresh Jewish cultural identity.' For her part, the British social anthropologist Sharon Macdonald writes of the *Purimspiel*: 'I am struck by another, disturbing, possible layer of inversion. If it is performed in February in a Catholic area, this is presumably literally Carnival. The hanging of Haman would, then, be an inversion of the atonement that it seems to perform.'

References

Barthes, Roland (1977) *Image-Music-Text*, London: Fontana/Collins.

Becker, Howard (2001) 'George Perec's Experiments in Social Description', *Ethnography*, vol. 2(1): 63–76.

Bellos, David (1995) *Georges Perec: A Life in Words*, London: The Harvill Press.

Benhabib, Seyla (2002) 'Citizens, Residents and Aliens in a Changing World: Political Membership in the Global Era', pp. 85–136 in Ulf Hedetoft and Mette Hjort (eds) *The Postnational Self: Belonging and Identity*, Minneapolis, London: University of Minnesota Press.

Blanchard, Tsvi (2004) 'Introduction', pp.vii-ix in Frédéric Brenner, *Diaspora: Homelands in Exile*, Volume 2: Voices, London: Bloomsbury.

Boyarin, Daniel (1992) *Storm from Paradise: The Politics of Jewish Memory*, Minneapolis: University of Minnesota Press.

Brenner, Frédéric (2004a) *Diaspora: Homelands in Exile*, Volume 1: Photographs, London: Bloomsbury.

Brenner, Frédéric (2004b) *Diaspora: Homelands in Exile*, Volume 2: Voices, London: Bloomsbury.

Carpentier, Alejo (1977) *Reasons of State*, London: Writers and Readers Publishing Cooperative.

Cavell, Stanley (2004) 'Departures', p.133 in Frédéric Brenner, *Diaspora: Homelands in Exile*, Volume 2: Voices, London: Bloomsbury.

Clifford, James (1997) *Routes: Travel and Translation in the Late Twentieth Century*, Cambridge, MA. and London: Harvard University Press.

Dayan, Daniel (2004a) 'The Rickshaw and God's Associate', p. 133 in Frédéric Brenner, *Diaspora: Homelands in Exile*, Volume 1: Photographs, London: Bloomsbury.

Dayan, Daniel (2004b) 'Passion Play in Tykocin', pp. 122–123 in Frédéric Brenner, *Diaspora: Homelands in Exile*, Volume 2: Voices, London: Bloomsbury.

Doyle, Roddy (1999) *A Star Called Henry*, London: Jonathan Cape.

Doyle, Roddy (2004) *Oh, Play that Thing*, London: Jonathan Cape.

Dylan, Bob (2004) *Chronicles*, Volume One, New York: Simon & Schuster.

Geertz, C. (1975), 'Thick Description: Towards an Interpretive Theory of Culture', pp. 3–30 in *The Interpretation of Cultures*, London: Hutchinson.

Ghosh, Amitav (2001) *The Glass Palace*, London: Harper Collins.

Hobsbawm, Eric (1995) *Age of Extremes: The Short Twentieth Century 1914–1991*, London: Michael Joseph.

Jaggi, Maya (2001) 'The Last Word', *The Guardian*, 21 December, www.http://books.guardian.co.uk.

Krauthausen, Ciro (2001) 'W.G. Sebald: Crecí en una familia posfascista alemana', *El País*, 14 July, http://www.elpais.es/suplementos/babelia.

Löfgren, Orvar (2002) 'The Nationalization of Anxiety: A History of Border Crossings', pp. 250–274 in Ulf Hedetoft and Mette Hjort (eds) *The Postnational Self: Belonging and Identity*, Minneapolis, London: University of Minnesota Press.

Macdonald, Sharon (2001) 'British Social Anthropology', pp. 60–79 in P. Atkinson, A Coffey, S. Delamont, J. Lofland and L. Lofland (eds) *The Handbook of Ethnography*, London/ Thousand Oaks: SAGE Publications.

Perec, Georges (1975) *W ou le souvenir d'enfance*, Paris: Editions Denoël.

Poupko, Yehiel (2004) 'Testimony', p.122 in Frédéric Brenner, *Diaspora: Homelands in Exile*, Volume 2: Voices, London: Bloomsbury.

Ruggiero, Vinceno (2003) *Crime in Literature: Sociology of Deviance and Fiction*, London: Verso.

Sachar, Howard M. (1994) *Farewell España: The World of the Sephardim Remembered*, New York: Vintage Books.

Schlesinger, Philip (2003) 'On Literary Ethnography: An Essay', pp.105–118 in Gunnar Liestøl, Bjarne Skov, Ove Solum (eds) *Mellom mediene: Helge Rønning 60 år*, Oslo: Unipub forlag.

Schlesinger, Philip (2004a) 'W. G. Sebald and the Condition of Exile', *Theory, Culture & Society*, vol. 21, no. 2, April, pp.43–67.

Schlesinger, Philip (2004b) 'On the Irrelevance of the Cold War: Some Reflections on the work of W. G. Sebald', pp.109–118 in Ruud Janssens and Rob Kroes (eds) *Post-Cold War Europe, Post-Cold War America*, Amsterdam: VU University Press.

Sebald, W. G. (1997) *The Emigrants*. London: The Harvill Press.

Sebald, W. G. (1998) *The Rings of Saturn*. London: The Harvill Press.

Sebald, W. G. (2000) *Vertigo*. London: The Harvill Press.

Sebald, W. G. (2001) *Austerlitz*. London: Penguin Books.

Sebald, W. G. (2003) *On the Natural History of Destruction*, London: Hamish Hamilton.

Sontag, Susan (1979) *On Photography*, Harmondsworth: Penguin.

Statewatch (2003a) 'EU: Buffer States and "Processing" Centres', *Statewatch: Monitoring the State and Civil Liberties in the UK and Europe*, vol. 13 no. 2, March–April, pp. 1–2.

Statewatch (2003b) 'Biometric ID Documents Herald a Global Identification system', *Statewatch: Monitoring the State and Civil Liberties in the UK and Europe*, vol. 13 no. 3/4, May–July, p.1.

Statewatch (2003c) 'UK Takes Lead on Surveillance of Passengers', *Statewatch: Monitoring the State and Civil Liberties in the UK and Europe*, vol. 13 no. 5, August–October, p. 1.

Statewatch (2004) 'EU: "Anti-terrorism" Legitimises Sweeping New "Internal Security" Complex', *Statewatch: Monitoring the State and Civil Liberties in the UK and Europe*, vol. 14 no. 5, August–October, p. 1.

Todorov, Tzvetan (1990) *Genres in Discourse*, Cambridge: Cambridge University Press.

Vargas Llosa, Mario (2001) *The Feast of the Goat*, London: Faber and Faber.

Williams, Raymond (1976) *Keywords: A Vocabulary of Culture and Society*, London: Fontana/Croom Helm.

Williams, Raymond (1977) *Marxism and Literature*, Oxford: Oxford University Press.

Imperialism, Self-inflicted? On the Americanizations of Television in Europe[1]

Jérôme Bourdon

Is there such an entity as 'European television'? Although politicians have used the syntagm to claim that 'European television' needed to be 'defended' and 'protected' from an American 'invasion', and although there are many institutions in Europe that claim to deal with or to represent European television, there is no such thing as far as viewers are concerned. There are no European programmes reaching a European audience across the continent or even a sizable part of it, there is no European station – unless the hybrid, alternately French and German *Arte* is rescued for the sake of the argument. There were, however, many attempts to create 'genuinely' European programmes and European stations, almost all of which resoundingly failed (Bourdon, 2007).

Some would claim that, although there is no European television in terms of content and audience, there is – or has been – a distinct legal and organizational model of television based on public service which addressed its audience as citizens first, consumers second, and promoted, in each national context, high-brow cultures based on history, literature and the arts. This model has been defined by contrast with the American, commercial model of television. It is now considered as under threat, and some would claim that it has fallen (Tracey, 1998). All agree that the golden age of public service organizations eager to inform and educate (first) and entertain (second), confident in their 'lofty ideal' (Avery, 1993, p. XIII), is over. However, historical research (for example, Bourdon, 1990, Palacio, 2001 Grasso, 1992, Hickethier,

1998) has shown that public service corporations were not the strongholds some retrospectively imagined and idealized. From historians' accounts, a sense of vulnerability emerges which can be analyzed in several ways. Public service organizations were almost constantly taken to task by politicians and threatened by radical reforms; they had to search for cultural legitimacy denied by intellectuals; they were divided along rival professional categories; and early on they had to embark on a search for the ever elusive audience as competition emerged, even within public service, and as, from 1956 to 1967, advertising was introduced almost everywhere.

Despite those reservations, it is easier to give substance to 'European television' as policy model than as programming practice. Nevertheless, this notion has weaknesses as well. First, it is normative as well as substantive. All too often, authors who describe public service also mean to extol it as the best way to defend certain values and to protect 'national European cultures' from Americanization or from 'cultural imperialism'. Many historians of European television, both professional and academic, concentrate on national territories and forget that major public service television operations were also tools of imperialism. Together with the defence of public service, the notion of Americanization has been used by European scholars with a nationalist touch. Eager to promote their own national culture through the defence of Europe, they easily forget that there are many forms of European imperialism beyond the American model.

All the Americas of Europeans

To complicate the picture further, it makes little sense to write in terms of relations from country to country (Lacoste, 1989, p. 242), not to mention continents. Countries are often reified as social actors (as are governments and television stations). They are social representations, and rarely can we imagine that a whole country wholeheartedly agrees on a policy. Before talking about 'national cultures', we have to understand how cultures travel and what they are made of. Framing cultures in terms of national cultures is often a strategic move, as the argument touches sensitive chords and allows the mobilization of supposedly common 'identities' while obfuscating differences and divergences. Television cultures, often described as national, will be analyzed here 'below' and 'above' the nation, as a set of professional and organizational practices within national cultures, but with many international interactions. I take a particular interest in groups of professionals that did not endorse the nationalistic cultural discourse opposing American cultures, 'Europeans' who applied for a different sense of identity than the one proposed by public television, and found no better symbolic resource for this than America – more precisely, than American television genres and formulas which could be adapted without provoking hostile or negative reactions. This is just an example of a wider phenomenon: the nation is always a weaker entity than many assume because it is conceptualized in different ways by people who claim to belong to the same nation.

The matter is also obscured by the fact that there are differences between the public political discourse, and more private discourses, in specific professional circles. Particularly, discourses on 'America' may differ. In a survey of French cinematographic culture (Bourdon, 2002), many respondents felt they should qualify their heavy consumption of American cinema with reservations about the need to defend 'French cinema' – which sounded very much like an 'imposition of problematics' in Bourdieu's terms. I am not sure whether they cared about 'French cinema' – but they cared to claim their support for French cinema, following the official discourse of French cultural nationalism, of which cinema is a stronghold. At the level of private cultural consumption, it is important to remember that few people actually embrace a full-fledged opposition to American culture and to American 'imperialism'. Everybody has been, at some point, a consumer of American culture, either through direct consumption or indirectly and often unknowingly through format adaptations. Opponents of Americanization might lament their countries being Americanized, but rarely do they reflect on the question of their own Americanization.

Everybody's cultural mix includes some amount of 'America'. This chapter could have been called 'All the Americas of Europe'. It is also about each European's own America. It is relevant to recall the presence of this diversity within nations which have had a long history of relations with America, where fascination and repulsion were often mixed, and where ambivalence, it seems, was fundamental, starting from Tocqueville's encounter. In the final analysis, it can be found among every opponent of 'Americanization' and 'American values'.

Imperialism, and self-inflicted imperialism

Ironically, then, this chapter is about imperialism, of the self-inflicted variety. By this I mean not that there is no such thing as media imperialism. Very early on, America public authorities conceived American television as a way to export American values. The international market of television was a very 'unpure' market bent to certain commercial-cultural-political aims. Segrave (1998, p. 33) gives numerous examples of American television suppliers being supported by the State Department, from a New York consultant in Germany in 1953 to a major distributor inviting 18 hosts to his Cincinnati headquarters in 1955. The USIA (United States Information Agency) encouraged both types of action in the Sixties, among developing nations and the Communist countries, but also in western Europe (ibid., p. 83). The most common tool (illegal in the US) for forcing exports on buyers was 'block-booking', also known as packaging or 'sales by volume'. This was used systematically in Third World countries. In order to purchase broadcasting rights for popular films or series, some stations had no choice but to buy less interesting, older fare. In Europe, this was more difficult to impose since public stations in particular were more reluctant to buy US products, and some countries (the UK and Germany) quickly developed powerful means of production. Until the Eighties, American exporters repeatedly complained about the reluctance of European stations to buy American fare, or their decisions to purchase small quantities or only parts of long-running series.

While exporters of fiction could rightly complain about European buyers' reluctance, other European media workers were busy importing American television, in a way which we would today call a transfer of 'formats' (but the word had not yet been adapted by professional European parlance). America was not imposed on them: they needed, admired, used, and imported American cultural forms. Within fragile public service stations (see above), a specific kind of Americanization took place. 'America' became a resource for professionals threatened from outside their organizations, and sometimes from inside. This 'America' was not always conceived, perceived and received in the same way. But, at a time when official public policy stressed the national character of public service (even for the only European private channel, ITV), something different was happening at the professional level. American television provided a source of inspiration, and sometimes a solution, to specific problems of programme elaboration.

America: models and counter-models

The initial context of television's development was certainly unfavourable to American imports or adaptations. It cannot be denied that, at least among politicians who contributed to the early debates on the launching of television, and of radio before that, America was forcefully rejected. 'Right from the start British engineers, legislators and civil servants looked at American broadcasting practices, and what they found and continued to find well into the age of television was not to their taste. The tone for this virtually permanent disapproval was already set by the Postmaster General in a reply to a Parliamentary question in April 1922, years before what we now call public service broadcasting came of age' (Hearst, 1992, p. 62). The UK is a key-country to our discussion. The country where commercial television was first introduced had the earlier and richer debates on broadcasting policy, in the course of which were described 'two models' of broadcasting: an American, commercial, competitive model, and a European, regulated, public service model. In the Fifties and the Sixties, this opposition would pervade all European debates, and later enter communications textbooks. Undoubtedly, the favourite model of the European was the British one, with the BBC as its synecdoche – while conveniently ignoring ITV, which is, not coincidently, a key station to the story we recount here. Apart from the BBC, alternative sources of inspiration were used, of course: debates and even formats circulated from France to Italy and Spain, and from Italy to France. But until the Seventies, television managers across Europe (in programming, engineering and audience appreciation) constantly refer to the BBC as *the* ideal. British influence was especially strong in Scandinavia. (In Sweden, television started in 1956 and was heavily influenced by the BBC. Swedish public television still broadcasts no advertising.)

To become a model for Europeans, Britain itself had to build the representation of the contrasting American and British models. This started at the time the BBC was set up: 'since the very beginning, the aim was to avoid chaos and the anticultural degenerations which were thought of as typical of US broadcasting' (Camporesi, 1994a, p. 267). Some comments acknowledged the 'variety', the 'energy', and the 'liveliness' of American television – whether those were linked to the American 'character', or, more dangerously, to competition (it would become a critical argument at the time of launching ITV). Before the 1954 bill, many journalists and some BBC officials visited the United States and came back, overall, with quite negative comments. At that time, a tone of permanent disapproval of American television policy was firmly set and could be generalized to most of what was published on television in Europe: the influence of advertisers, the transformation of presenters into show-business stars (especially in news), the cultural poverty, were constantly under attack (for France, see Bourdon, 1989). Even the campaign for the introduction of commercial television in Britain had strongly anti-American tones: commercial television, it was said, had to be introduced to prevent American domination.

Television with advertising would be done 'the British way'. For aren't the British 'a much more mature and sophisticated people' (Camporesi, 1994a, p. 279), as a British conservative MP would ask in 1952. This reflected a strongly felt and ancient European prejudice towards the United States (Kroes, 1996).

America did have its supporters. We will soon find examples inside television. Outside, four nascent 'professions' had long relied on some sort of American expertise: journalism, advertising, commercial radio broadcasting, and pollsters. The British press had very early on been affected by a process of Americanization which reached a peak at the end of the nineteenth century. At that time, many European countries adapted models of American newspapers (sometimes through the UK). Amidst reluctance from the elite press, newspapermen travelled to England and the United States. In 1903, a French journalist wrote: 'the United States is the Holy Land of journalism' (Chalaby, 1996, p. 317). Advertising was the other major field of influence. Some advertisers went on US pilgrimages at the end of the nineteenth century, and were glad to find a country where their profession, viewed with hostility within European culture, was given consideration (for France, see Pope, 1978). The first advertising agencies were key places for adaptation of American concepts. This influence could filter out of the advertising milieu, especially through a medium like commercial radio, in the rare countries where private radio was tolerated or accepted. In France particularly, advertising financed radio before the Second World War, and commercial 'radio *périphériques*' continued broadcasting after the war, despite the official State broadcasting monopoly. In the Fifties, Louis Merlin, ad-man and radio producer, tried to pioneer commercial television. In 1946, he published a book entitled *Télévision, Capitale Hollywood* – which quickly sank into oblivion. Fourthly and finally, the first social scientists who took an interest in public opinion measurement were attracted by the United States, where the technique of opinion polls was born, and tried to adapt it to their own countries, sometimes before the Second World War (Blondiaux, 1998).

As for television, it viewed itself, in the Fifties, as firmly located within the public service tradition and as a tool to promote national culture. Everywhere, national versions of the famous three-pronged missions ('inform, cultivate and entertain') were formulated and developed. Entertainment came third and last, which was a way of saying that television was not commercial television. The only genre that was, early on, at the centre of most discussions of American influence on television is fiction, that is popular television series which were imported into Europe in the Fifties and the Sixties. *The Untouchables,* for example, triggered debates regarding content (violence especially, in the US), but also on the necessity of producing local, attractive popular fiction.

In most European countries, there was strong resistance to adapting American game-shows formats. This resistance was all the more remarkable since these formats offered a virtually free source of programme ideas. In those days, 'plagiarism was rife', and most 'borrowing' took place without payment. When it happened at all, 'early payments for the use of program ideas were *ad hoc* and more in the nature of a courtesy to the original producer or owner' (Moran, 1998, p. 18). It is only in the Seventies that a license fee system would emerge. And it was only in the UK that, before that period, we have any reference at all of payment to an American producer – which can easily be explained by the close links between media industries of both countries. In practice, it seems that most European producers simply travelled to the States; some visited installations and studios, some simply watched TV, liked an idea, a visual detail, and freely picked it up. We will find many examples of these 'American television pilgrimages', but nowhere as many as in England and Spain. This shows that the question of American imports is not connected so much to pre-existing 'cultural affinities' (Heinderyckx, 1993. Both 'Latin' and 'Anglo-Saxon' countries were early and heavy importers of American fare. The explanation lies, very simply, in the early and massive introduction of advertising.

England and Spain: the influence of advertising

As opposed to the rest of Europe, England and Spain had one strong common organizational feature: in each country, one television channel was entirely financed by advertising. The already Americanized advertising culture had a chance to influence the new medium of television. In England, ITV started broadcasting in 1955. In Spain, television officially started in 1956, as a department of the direction of broadcasting, itself attached to the ministry of Information – but from the start most of the money came from advertising.

In England, the nascent ITV relied heavily on American light entertainment to ensure its success, be it situation comedies or quizzes, which the press called 'give away programmes'. The UK already had the richest history of American media format adaptations, especially in radio (Camporesi, 1994b). In January 1957 there were no less than ten quizzes broadcast each week on ITV (Sendall, 1982, p. 348), most of them copies of American formats, including *The 64.000 Dollar Question*, *Spot the Tune, Double Your Money*, *Twenty One* (faithfully memorialized in Robert Redford's movie *Quiz Show*), and *Beat the Clock*. They were then prime time programmes, as in the United States. They came under heavy criticism from the Independent Television Authority which supervised ITV at the time, and from the press, but the bags of post the company received told a different story: thousands of people wanted to participate. The content of quizzes, especially the value of prizes, not to mention the programming itself, was submitted to more stringent rules (especially after rumours of scandal and rigging up around Granada's *Twenty One* in 1959, the year of the quiz-shows scandal in the USA). The UK became the major European programmer of quizzes and would remain so for a long time (*Eurodience*, 1988).

Under the pressure of competition, the BBC also resorted to game-show formats. In the UK, unlike elsewhere in Europe, time slots (even in prime time) were short – half an hour or one hour, and the American formats could be more faithfully copied. England refused to 'freeze' programme schedules for long periods of time. Some formulas were used only for one season, shelved, and then reused again – which can be analyzed as a way of underlining British television's relative freedom from competitive constraints.

In Spain, the need to secure advertisers' money quickly created tensions with public service ideals. Some programmes (*La Hora Philips* or *Festival Marconi*) were sponsored, with presenters and emcees promoting the product in the American way. Quizzes were imported early on but immediately subjected to an adaptation process that Grasso (1992, p. 62) has called, in the Italian context, 'theatrical dilatation'. They were lengthened and sometimes incorporated into longer programmes. In 1959, TVE introduced a prime time variety programme called *Club Miramar*, which included variety, 'concursos' (games), and interviews. *Club Miramar* was followed by *X-O da dinero*, the Spanish Version of the American *Tic-Tac Dough*, which Jorge Leman (pseudonym of the advertising professional Jorge Garriga Puig) had 'discovered in one of his frequent professional trips' to the United States (Baget Herms, 1993, p. 61). Nestlé Spain, whose head of advertising was the same Jorge Leman, sponsored the programme.

Italy and Germany: quizzes turn spectacular

Did professionals in other national contexts look to America for their inspiration? Undoubtedly – and again, the trend was established long before television. Italy was the continental democracy most attracted to the United States, evidenced in part by the many pilgrimages made by Italian television professionals to America (Grasso, 1992). Sergio Pugliese, the first director of Italian television (1954–1962) and a drama writer who stressed the cultural missions of the medium, visited American television shortly after taking office. It is no coincidence that the most famous presenter of Italian television, Mike Bongiorno (literally: Mike Good Day) was born in New York of an Italian mother and an American father. Before his television career, he worked for a radio programme aptly called *Progresso Italo-Americano*.

American examples and sources of inspiration abound in the early history of Italian television. One year after the start of broadcasting, one can find an adaptation of an American quiz: *Duocento al secondo* (*A Dollar a Second*), expanded to one hour. Broadcast from June to September 1955, it was interrupted after a deluge of critiques about the punishments meted out on participants, which were considered 'offensive to human dignity' (another sensitive point among critiques of game and reality shows) (Grosso, 1992, p. 59). American models influenced early variety shows. 'The influence of Ed Sullivan...is to be found in the luxurious and flashy sets and in the prestige of the stars which were the main attraction of the shows' (Grasso, 1998, p. 47). From 1955 to 1959 (with some interruptions), at usually at 21.30, *Un Due Tre* was an hour and a quarter long musical variety show that signalled the 'passage from the review theatre to the television review' (Grosso, 1992, p. 48), with an obvious proximity with NBC's *Your Show of Shows*. On May 20, 1958, Italy became probably the only country that broadcast an original version of an American variety show, *The Perry Como Show*, with subtitles and without success. Antonello Falqui, the producer of *Studio Uno* (1961–1967), returned from America with the notion of a less theatrical and more open space where the apparatus of television production was made visible to the viewer (Grasso, 1992, p. 138).

German television (Erlinger & Foltin, 1994) shared the tendency to transform quizzes into large theatrical shows. The stereotypes about German culture are ill-fitted to describe this type of programme, where screams and laughter can be heard in spectacular settings. In this area – as in news – early German television had strong professional links with American networks.

In the beginning, Peter Frankenfeld and H.J. Kulenkampff dominated the field of television entertainment. The first had worked for US radio in Germany. The second went on numerous trips to the United States. They launched a variety of formulas such as, in 1954, *1 :0 für Sie*, adapted from the US radio series called *People are Funny* – a mixture of variety and games that was later adapted in Sweden.

As in Britain, the introduction of advertising on television (starting with Bayerishes Rundfunk in 1956) prompted a multiplicity of game shows around advertising slots (*Tic Tac Dough* became *Tik Tak Quiz*, broadcast from 1958 to 1967). *Alles oder Nichts*, the German version of *The 64.000 Dollars Question*, was broadcast from early in the Sixties until 1988. Starting in 1958, *Was bin ich?* (*What's my Line?*) remained a success until 1989. *Hätten sie es gewusst* (until 1988) was inspired by the American *Twenty One*. Game shows were transformed into longer, prime time or early prime time formats, sometimes broadcast on a monthly basis (as they still are today on German public television). This weekly or monthly rhythm allowed such shows to last much longer than in the rest of Europe. German television developed a rich culture of game and variety shows, exporting its own formats, such as *Wetten Dass* (ZDF, 1981) to the UK, Italy and, a rare feat before the age of reality-shows, the US.

French reluctance, and the emphasis on culture

In France, the overall volume (and the American adaptations) of game shows has been the most modest. Producers and directors worked in an atmosphere hostile to game shows. A small group of still-active producers (the most famous being Jacques Antoine and Pierre Bellemare) initiated some original formats: *Télé-Match*, which became *la Tête et les Jambes* (The Head and the Legs) in 1956, was based on the complementary performances of an 'intellectual' in the studio, 'the head', and of a sports performer in the gym, 'the legs'. It was adapted by Spanish and Italian television in the late Fifties and licensed to American television in the Seventies. But, even in its game shows, French TV put the accent on 'culture'. The two longest running game shows were based on knowledge of vocabulary and algebra (*Des Chiffres et des Lettres*, from 1972 onwards, which became *Countdown* on Channel 4), and on *cinéphilie: Monsieur Cinéma* lasted from 1966 until the mid-Eighties under various titles.

The only known case of French pilgrimage to the US before the Seventies is that of the most famous variety producers, the Carpentiers, who went to the States to survey the production of *The Perry Como Show*. France adapted at least one famous American format: *Candid Camera*. The show (CBS, 1960–1967) had started its European career in Italy as *Specchio Segreto* (RAI1, 1965, eight programmes of one hour). As in the Spanish version (*Objectivo Indiscreto*) that started the same year, there was an emphasis on the moral conclusion. In Spain, the programme was controversial because it supposedly exploited 'innocent' victims by ridiculing them. In some cases, original footage from the United States was imported. In France, the programme started in 1966 under the title *La camera invisible*. A television director had just seen the programme in the course of a pleasure trip to the US. It became a huge 'French' success. A professional trip to the United States, an isolated adaptation: that was the most that could be said for American influence at the time. Unlike all major European countries, French television did not adapt the American *This is Your Life* (1952–1961) in the Sixties, but only in the Nineties and without success.

Most European game shows of the Fifties and the Sixties have one thing in common: the tendency to stress 'serious knowledge', not trivia, to remain faithful to an educational mission, and to propose more modest prizes than their US counterparts, as in the 1950s Swedish adaptation of *The 64,000 Dollar Question* (Kleberg, 1996, p. 194). The 'American' format became quite 'Swedish', appearing as a totally *sui generis* achievement of Swedish public television, especially when one of its first candidates became a young star in Swedish culture, as happened in other countries as well.

News before the anchorman

Entertainment genres may have been considered necessary to attract audiences, but they were not central to the sacred missions of public service and were viewed with contempt by drama producers and respectable critiques. We now move to a more legitimate genre, information – and more specifically, to news. An Americanization of newscasts occurred in the Seventies, and unlike the previous entertainment-based Americanization, the US was publicly vindicated as a model of good television. As a point of reference, consider news formats in the Fifties and the Sixties. The variety was incredible. The distinction we are now familiar with between 'news' and 'current affairs' was being drawn and discussed under different rubrics (in France, between 'journal' and 'magazine', in Spain, between '*telediarios*' and '*programmas de actualidad*'). The way news should be presented, by whom, with what proportion of pictures, was the topic of a debate that would culminate in the mid-Seventies – with the advent of the anchorman (not yet an anchorperson at the time).

In the beginning, the lonely anchorman had yet to run the show – and the word 'show' is most inappropriate to describe the news. In some places (Italy, Spain), speakers (not journalists) read the news from paper (like they used to do on radio). The journalist's job was to find the news and prepare it, but not to get into a relationship with the audience. A universal trait was the deliberate rejection of the star-system, which was understood as jeopardizing the news' 'neutrality' and 'objectivity'. The BBC exemplified this television Puritanism. 'On 4 September 1955, less than three weeks before ITN's first transmission, the BBC had shown its newsreaders' faces on screen for the first time; even then, however, the newsreaders were not named, for fear that this would affect the bulletin's impartiality' (Wheen, 1985, p. 72). ITN played a major part in 'americanizing' the format of TV news. The news was not read by newsreaders but by newscasters, journalists who would write their own scripts and would also act as interviewers and reporters. For the first time, TV viewers were invited to identify the news with a personality (not only to an expressionless character reading a text).

In France, journalists had been shown on the screen reading the news earlier than on ITV. But the context was quite different. The first 'journal *télévisé*' (called the 'first in the world' by his creator, seemingly unaware of the launch of CBS evening news on May 3, 1948), was broadcast on June 1949. A small group of producers and newsreel cameramen (using their own personal cameras) put it together with very few resources but much enthusiasm. Unfortunately, the programme was not judged worthy of much attention, whether by politicians or fellow journalists. Many of its 'founding fathers' were not journalists. Until 1954, an improvized combination of footage made available by foreign embassies and original (silent) 16-mm footage, was commented upon live. In November 1954, journalists were shown reading the news in the studio, and until 1971, there were many presenters even during the same edition. From 1963 to 1965, the 'formule' (the Americanism term 'format' had not entered the French language) of the '*presentateur-vedette*' (the presenter-star) was tried. A few presenters became stars, which television heads did not like, since promoting a specific personality was not considered to be journalism. In 1965, 'the image' was officially given priority (a recurring motive in the history of news). The '*nouvelle formule*' was tried, that is, the news returned to the system of several newscasters during the same evening edition, with a higher proportion of stories.

The professional journalistic reference: from the UK to the United States?

Up until the early Seventies, what was the reference of good journalism for television journalists in Europe? In Southern Europe, where independence from political control was, to say the least, fragile, the BBC was considered as the model of professionalism and independence, and was often quoted as such in professional documents and parliamentary reports. However, the United States was present as a reference as well, at least in some countries, and for some specific genres or professional practices. In Italy, at least in current affairs programmes, the presence of the correspondent at the centre of his story was introduced in the magazine *TV7* (1963–1971), following the 'American model' (Grosso, 1992, p. 174). But the other American genre that mattered was the *Meet the Press* format, introduced in France in 1965 as *Face-à-face*. Interestingly enough, a supposedly 'aggressive' type of journalism was not deemed British (although it was present in England at the time). We can surmise that a shift occurred in the mid-Sixties: at some point, the reference to modern journalism moved from current affairs programmes (a flagship of British television) to political debates and to the main news – and, at the same time, from the UK to the United States.

A particularly interesting European figure here is the 'New York correspondent'. Jacques Sallebert for France, Jose Hermida for TVE, and Ruggiero Orlando for RAI embodied modernity for professionals. Becoming the New York correspondent was a promotion, and frequently a major station in a brilliant career. In collective memory, their voice was associated to major international-American media events such as John Kennedy's funeral or the man's first steps on the moon.

Competition: the anchorman comes of age

In the early Seventies most European newscasts adopted the format now familiar to world viewers. During a half hour, a single, familiar newscaster (more rarely a couple) tells the major 'stories', 'launches' reports, and interviews guests. The single newscaster was present from the start in the US, but for short newscasts. Only in 1963 did American networks begin broadcasting 30 minutes newscasts, at 18h30. This was meant to help them regain prestige after the quiz shows scandal and the many critiques of the networks' cultural 'mediocrity'. In Britain, the editor of ITV news since 1956, Geoffrey Cox (Cox, 1995) lobbied relentlessly until the ITA allowed him to launch *News at Ten* in 1967, with two newscasters, following the then 'new' American habit of the time. The BBC followed suit, and lengthened its own bulletin into a series of 'reporters' stories' around the personality of the presenter.

In the rest of Europe during the years 1968 – 1974, internal reform and especially internal competition was on the agenda of public service broadcasting. This was directly related to the transformation of newscasts. In France, the second channel was created in 1964, but serious competition was only introduced in 1969. The newscasts of both of the main two channels have been broadcast at the same time ever since. In 1971, the second channel introduced the '*présentateur* unique'. After the 1974 reform, which split television into three separate channels, the first channel introduced a single newscaster for each of its three editions: mid-day, evening, and late night. Joseph Pasteur, the first 'présentateur unique', and Roger Gicquel, the most popular one in the Seventies, both made a trip to America before starting and met their model Walter Cronkite.

In Italy, the much criticized and ridiculed RAI newscast was transformed in 1968. 'For the first time, it was led by journalists and not by speakers'. They were 'four voices, one leader, one for foreign news, one for domestic news and one for sport' (Veltroni, 1992, p. 265). In 1975, the new reform that, as in France, gave more independence to the (two) television channels, also led to the establishment of new more personalized formulas. In Spain, the second channel started broadcasting its first newscasts in 1965. Journalists started reading their own text in 1967; in 1974 each journal was edited and presented by a single anchorman. In Germany (Ludes et al,. 1994), public channels were independent organizations from the start. ARD's *Tageschau* enjoyed a monopoly from its inception in 1956 until ZDF started broadcasting in 1963. It was broadcast at 20h. In spite of an official agreement between both organizations, ZDF put some pressure on its rival by broadcasting its own newscast at 19h50 (ten minutes before ARD's), and then at 19h30. ARD fought back by airing its regional broadcasts at 19h25. In 1969, ZDF lengthened its news to 25 minutes, while ARD has remained faithful, to-date, to its brief 15-minute newscast. Several formula changes occurred. From the beginning, ZDF looked to the US for professional inspiration. Its news was (relatively) more spectacular. In 1965, the editor became newscaster of the news (and not simply commentator of stories). In 1973, finally, ZDF introduced the 'stake out': the correspondents spoke to the audience within their stories, and their names appeared on the screen.

The single evening star was not universally accepted. It drew the fire of critics who feared the power of this new kind of celebrity. In 1978, TVE, after three years of applying the Cronkite model, went back to a 'main newscaster' assisted by less prominent 'secondary newscasters'. Professionals call this a return to a 'more British' model (Baget Herms, 1993). It would take ten years for the anchorman to be accepted as the appropriate model for television news. Unlike the US, not all countries felt that the presenters should also be the editors. In France, it was only in 1981 that this supreme mark of independence was given to an anchorman – actually an anchorwoman, with some vaunted professional experience in the US.

Deregulation, and the New Wave

'Self-inflicted imperialism' depends less on national cultures, or on national characters, than on the internal dynamics and the structural characteristics of each television system (especially the introduction of competition and advertising, which encourages the invention, import or adaptation of commercial formats). For the most part, politicians, heads of public television, and 'artists' (that is, television directors) within public television loathed the idea of Americanization. Against this background, light entertainment producers, followed by journalists, discreetly found inspiration in the US, without being committed to the American system as such. All this went largely unnoticed by audiences, including many professionals, because, as Jean Chalaby (1996, p.323) has noted about the print press, 'the export of discursive practices is much less visible than the export of actual movies and television series'. In those early days of public service, professionals in all countries boasted of their ability to domesticate the American formats, just as the British had done when ITV started. In Italy, RAI made intense efforts to Italianize American shows. It has even been said that in those early years, 'when television aspired to a faraway American dream, it most clearly found its originality' (Buttafava, 1980).

The deregulation of the Eighties had a decisive impact on European television. Deregulation, conceived by many politicians as a way of imitating American television, made the job of the old 'imperialists' easier. Deregulation gave a chance to American exporters to add Europe to their list of conquered territories. It provoked a turn to Europe: 'by 1987 the top eight markets for American television product accounted for 80 percent of US total offshore sales; five of them were in Europe – Britain, France Germany, Italy and Scandinavia – with the others being Australia, Canada and Japan. Latin America's share of US business stood at 15 percent of the total, down considerably from years earlier' (Segrave, 1998, p. 174). Europe became the new 'promised land', especially as new commercial stations were willing to buy old series and formats which provided their early bread and butter.

At this juncture, professionalization became a form, and a positive one, of Americanization. While until the Seventies professionals claimed they could 'disguise' and domesticate American ideas, new television professionals of the private sector boast exactly the opposite: the ability to be as 'professional' as the Americans, to respect the format. To quote a Canal Plus general manager, representative of the new generation of '*Américanophiles*': 'It took the pressures of audience results and experience in working closely and methodically with producers in English-speaking countries for this to come about. The curtain has therefore fallen on former practices of taking over famous formats, for the most part American, and liberally toning them down on the excuse that they had to be adapted. It was all very good-natured, inexpensive and empirical in the good old French tradition' (*Eurodience*, 1988).

In the Eighties, there was much more respect for original American formats, and, beyond that, for the American system as a whole. America had been partly rehabilitated by politicians; neoliberalism has become the model for policy at large, even when the official discourse went on defending public service and the welfare state. In television in particular, Americanization was conscious, systematic, and fast. Professionals knew it was taking place, and not only in their countries; they also knew that, in a state of acute competition, they had to find the right (American) resources as early as possible. Even the (rather rare) fierce opponents of American television agreed on its dynamism, on the quality of (some of) its popular fiction, of its newly vaunted 'professionalism'.

In Europe, the symbol of this transformation was the arrival of *The Price is Right*, created in 1956 by Goodson-Todman for NBC. It embodied everything public service television once reproved: emphasis on money, minimum 'knowledge' with a huge amount of luck. It started being programmed by Berlusconi's Italia Uno in 1983 (*OK il Prezzo e giusto*), ITV in 1984, TF1 in 1988 (*Le Juste Prix*), and RTL (*Der Preiss ist heiss*), among many others. *The Wheel of Fortune* (created in 1975 by Merv Griffin, and a success on US local stations in 1983) became a symbol of the change in programming when the soon-to-be-privatized first channel bought the rights through the advertising agency Lintas in 1986, and broadcast it with success in prime time. In Italy, it was first broadcast on the now defunct commercial station Odeon TV, until Berlusconi bought it in 1989. It was first broadcast as a long weekly show on Canale 5 on Sunday evenings, then as a daily programme from 1991 onwards.

The anchorman has also spread everywhere, although in northern Europe public service broadcasters still insist on the moderate, not committed style of their newscasters. In Germany, ARD has long remained the last stronghold of journalistic Puritanism: in 2000, its main evening news edition was still led by seven newscasters who were '*Sprecher*' (speakers), not journalists. But, in the Eighties, private channel Sat 1 launched an evening news edition with a young journalist, Ulrich Meyer, with the avowed aim of creating a popular news star. In France, Italy, and Spain, the celebrity of the newscaster had been accepted earlier, even though it is still occasionally criticized.

Different notions of Americanization

The time has come to recapitulate the semantic transformations undergone by the elusive term 'Americanization'. Some caveats are in order here. The term supposes, among other things, that 'American television' is something stable and unchanging in the course of the years. Without going that far, we can assume that, relatively speaking, American television was actually more stable in its programming foundations than European television. Second, Americanization is often linked to the transfer of 'values' from one society to another. This chapter is about television and about television professions, programmes and policies. It claims key processes have been at work and so far ignored. It makes no direct claim regarding the impact American programmes might have had on European cultures at large.

We started by recalling that Americanization could be, and has been, about **imperialism**. Although in Europe, we witnessed another form of discrete Americanization that I would call Americanization as **popularization** (elaborating content adapted to mass audiences, which, for European television of the Fifties often meant copying or reinventing American television, just like the European press did in the late nineteenth century). This form of Americanization can go unnoticed by audiences. For professionals, it might take on another meaning: Americanization can be seen by some as a form of **professionalization**: that is, using America as a resource among professional milieus, to reach a certain status, for example, as anchorman or producer-host. With deregulation, Americanization triumphed as **commercialization**: media professionals used American contents, consciously and conscientiously, in order to draw viewers who were now addressed as consumers into a new commercial and competitive environment. Although deregulation gave this process a decisive momentum, it happened early in England and Spain. But deregulation was also about Americanization as **policy**: the conscious (albeit not necessarily faithful and successful) replication of American models for reorganizing television, from State monopoly to a much more lightly regulated competitive commercial medium.

America as professionalization is at the centre of this paper. It is crucial because this is the first area where Americanization was viewed as a positive aim to be reached. Without discarding the media imperialism thesis, many other factors should be taken into account especially in the American-European context. In addition, one should analyze the relationship between the different notions of Americanization. All these notions are to some extent valid. All are connected. In particular, a clear causal relation links policy to professional culture. Creating a commercial television station (policy) is the best way to encourage popularization (through commercialization) and professionalization, both relying on American resources. It also makes traditional imperialism less necessary since it reinforces 'self-inflicted imperialism'.

Processes of cultural influence take place in many different ways. Recent analysis has moved away from notions of media imperialism and highlighted a more multidimensional, networked, fluid globalization. It has also focused on reception and on the variety of interpretations that viewers can draw from the same content. However, between the alleged freedom of receivers and the alleged powers of imperialists, there is a crucial missing link. Reception can only be explained by taking into account 'the mediating role of gatekeepers in creating a reception context' (Cunningham and Jacka, 1997, p. 308). We still have much to learn about the European gatekeepers of Americanization.

Note

1. An earlier, shorter version of this paper was published in French in *Reseaux*, 19/107, 201, and in Italian in *Contemporaneo*, 4/1, 2001.

References

Avery, R. K., Ed. (1993). *Public Service Broadcasting in a Multichannel Environment*. New York, Longman.

Baget Herms, J.M. (1993) *Historia de la Television en Espana (1956–1975)*. Barcelona, Feed-Back.

Blondiaux, L. (1998). *La fabrique de l'opinion. Une histoire sociale des sondages*. Paris: Seuil.

Blumler, J.G. (Ed.) (1992) *Television and the Public Interest*. London: Sage.

Bourdon, J. (1989) 'Les détours de la séduction : les Etats-Unis et les professionnels de la télévision française', pp. 171–187 in C.J. Bertrand and F. Bordat (eds), *Les médias américains en France*. Paris: Belin.

Bourdon, J. (1990) *Histoire de la télévision sous de Gaulle*. Paris: Anthropos & Institut national de l'audiovisuel.

Bourdon, J. (2002) 'Spectator culture as popular culture. Life-stories of French moviegoers', *Comparisons, A international journal of comparative literature, Vol 2/2*.

Bourdon, J. (2007) 'Unhappy Engineers of the European Soul.The EBU and the Woes of Pan-European Television', *International Communication Gazette, Vol. 69,* No. 3, 263–280.

Buttafava, G. (1980) 'Un sogno americano. Quiz e riviste TV negli anni 1950', in Buttafava, G., Grasso, A. Lombezzi, M. Sanguineti, T. *American way of television. Le origini della TV in Italia*. Firenze: Sansoni.

Camporesi, V. (1994a) 'There are no kangaroos in Kent. The American 'model' and the introduction of commercial television in Britain, 1940–1954', pp. 266–282 in D. Ellwood, R. Kroes (ed.), *Hollywood in Europe. Experiences of a Cultural Hegemony*. Amsterdam: VU University Press, 1994.

Camporesi, V. (1994b) 'The BBC and American Broadcasting, 1922–55', *Media, Culture and Society 16-4*: 625–639.

Chalaby, J. K. (1996). 'Journalism as an Anglo-American Invention. A Comparison of the Development of French and Anglo-American Journalism, 1830s–1920s', *European Journal of Communication, Vol 11(3)*: 303–326.

Cox, G. (1995) *Pioneering television news*. London: John Libbey.

Cunningham, S. and Jacka, E. (1997). Neighbourly relations? Cross-cultural reception analysis and Australian soaps in Britain. In Annabelle Sreberny-Mohammadi and al. (ed.) *Media in Global Context. A Reader* (pp. 299–310). London: Arnold.

Erlinger, H.D. & Foltin, H.F. (1994) (eds). *Geschichte des Fernsehens in der Bundesrepublik Deutschland, Band 4. Unterhaltung, Werbung und Zielgruppenprogramme*. München: Wilhem Fink Verlag.

Eurodience, European newsletter on programmes and audiences (1988), no 12, July, and no 13, September (Paris, Institut National de l'Audiovisuel).

Grasso, A. (1998) 'Italie: les variétés se meurent, vive les variétés', *Dossiers de l'audiovisuel*, mars-avril: 45–48 (Paris, Institut national de l'audiovisuel).

Grasso, A. (1992) *Storia della televisione italiana*. Milano, Garzanti (new update edition 2000)

Hearst, S. (1992) 'Broadcasting Regulation in Britain', pp. 61–79 in J. G. Blumler (ed.). *Television and the Public Interest*. London: Sage.

Heinderyckx, F. (1993) 'Television News Programmes in Western Europe: A Comparative Study', *European Journal of Communication, Vol. 8*: 425–450.

Hickethier, K. (1998). *Geschichte des deutschen Fernsehens*. Stuttgart Weimar, Metzler.

Kleberg, M. (1996) 'The history of Swedish television. Three stages', pp. 182–207 in I. Bondebjerg, F. Bono, Television in Scandinavia. History, Esthetics and Politics. London: University of Lutton Press.

Kroes, R. (1996). If you've seen one, you've seen the mall. European and American Mass Culture. Urbana: University of Illinois Press.

Lacoste, Y. (1989). Géographie du sous-développement. Géopolitique d'une crise. Paris: PUF.

Ludes, P., Schumacher, H., Zimmermann, P. (1994) (eds) Geschichte des Fernsehens in der Bundesrepublik Deutschland, Band 3. Informations und Dokumentarsendungen: Nachrichten Sendungen. München: Wilhem Fink Verlag.

Moran, A. (1998) Copycat TV. Globalisation, Program Formats and Cultural Identity. London: University of Lutton Press.

Palacio, M. (2001) Historia de la televisión en España. Madrid: Editorial Gedisa.

Pope, D. (1978) 'French Advertising Men and the American Promised Land', Historical Reflections, V: 117–139.

Segrave, K. (1998). American Television Abroad. Hollywood's Attempt to Dominate World Television. Jefferson, North Carolina and London: McFarland.

Sendall B. (1982) Independent Television in Britain, Vol 1., Origin and Foundation, 1946–1962. London: Macmillan.

Tracey, M. (1998) *The Decline and Fall of Public Service Broadcasting*. Oxford: Oxford University Press.

Veltroni, W. (1992) *I programmi che hanno cambiato l'Italia. Quarant'anni di televisione*. Milano: Feltrinelli.

Wheen, F. (1985) *Television. A History*. London: Century.

Media and Cultural Diversity in Europe

Kevin Robins

Introduction

Until quite recently, what prevailed in Europe was the system of public service broadcasting, involving the provision of mixed programming – with strict controls on the amount of foreign material shown – on national channels available to all. The principle that governed the regulation of broadcasting was that of national 'public interest'. Broadcasting should contribute to the public and political life of the nation; in the words of the BBC's first Director General, John Reith, it should serve as 'the integrator of democracy'. Broadcasting was also intended to help in constructing a sense of national unity. In the earliest days of the BBC, the medium of radio was consciously employed 'to forge a link between the dispersed and disparate listeners and the symbolic heartland of national life' (Cardiff and Scannell, 1987, p. 157). In the post-war years, it was broadcasting that became the central mechanism for constructing this collective life and culture of the nation. In succession, radio and television have 'brought into being a culture in common to whole populations and a shared public life of a quite new kind' (Scannell 1989, p. 138). Historically, then, broadcasting assumed a dual role, serving both as the political public sphere of the nation state, and as the focus for national cultural identification. We can say that broadcasting has been one of the key institutions through which listeners and viewers have come to imagine themselves as members of the national community.

Over the past twenty years or so, however, things have changed, and changed in quite significant ways. From the mid-1980s, dramatic upheavals took place in the media industries, laying the basis for what must be seen as a new kind of media order. Two factors have been identified as being particularly significant in this transformation. First was the decisive shift in media regulatory principles: from regulation in the national public interest to a new regulatory regime – sometimes erroneously described as 'deregulation' – primarily driven by economic and entrepreneurial imperatives. Second was the proliferation of new, or alternative, distribution technologies, and particularly satellite television, which made it possible, maybe inevitable, for new transnational broadcasting systems to develop – bringing about, as a consequence, the formation of new transnational and global audio-visual markets. Driving these developments were new commercial and entrepreneurial ambitions in the media sector. And what was particularly significant here was the strong expansionist tendency at work in these ambitions, pushing all the time toward the construction of enlarged audio-visual spaces and markets. The objective and the great ideal in the new order – among media entrepreneurs and policy-makers alike – was 'the free flow of television'. The imperative was to break down the old boundaries and frontiers of national communities, which had come to be seen as restricting the free flow of products and services in communications markets. There was the tendency for the new audio-visual spaces to become detached from the symbolic spaces of national communities and cultures.

Discussions of these developments have tended to be seen in terms of the shift from one historical epoch or era to another – the transition from the public service era to that of global markets. In this metaphor of epochal shift, there is a tendency to overemphasize the contrast between the two phases, and also to oversimplify the nature of each phase. What I want to suggest is the use of a different metaphor to grasp the nature of the transformations that have been occurring. I would suggest that change is more akin the process of geological layering. What has happened is that the new audio-visual spaces and markets have come to settle across the old national landscape. Public service broadcasting continues to exist at the same time that new kinds of audio-visual concerns have come into existence. Also important to emphasize, I think, is that both 'public service' and 'global' are fluid and changing categories. In Europe in the 1990s, for example, the idea of public service shifted in important ways to include provision of programming for minorities and also the recognition of cultural rights in the European regions. We should be clear as well that global broadcasting has developed in such a way as to include transnational and diasporic broadcasters like Roj TV (formerly MED TV) and Al-Jazeera, as well as giants like Disney and Time Warner. If we consider the European continent now, what should be apparent is the extreme diversity and complexity of audio-visual spaces, national, local-regional, and transnational. Viewers may tune into the services of public service providers like RAI, ZDF, Welsh or Basque channels, CNN or Sky, and also ZeeTV or TRT. And through this process the nature of the European cultural and political geography is being recast.

Contemporary developments in media industries and cultures are crucial for contemporary Europe. If public service broadcasting was central to the institution of national cultures and communities, we may argue that the new broadcasting culture must be central to the imagination of the new Europe that is coming into existence. In the following discussion, I want to consider the significance of the new media culture with respect to cultural diversity in the new Europe. This could be a vast subject in itself, so I will confine my discussion to one core issue. This concerns the imperative to effect some kind of distancing from the national imagination and the national paradigm if we are to develop a European media policy that is sensitive to the new cultural diversity of the continent. In the 1990s, the European Commission struggled with the question of diversity within the framework of a policy primarily concerned with the need to create an expanded, pan-European market ('television without frontiers'). But diversity was seen in rather limited terms - in terms of the Europe of nation states and/or the Europe of the regions. The *imaginaire* remained essentially national. And it was an *imaginaire* that did not address - because it could not recognize - the actual complexity and diversity of Europe. The national imagination stood in the way of moving cultural thinking and policy forward in order that it might accommodate the realities of the new Europe.

In order to pursue my argument concerning the national imaginary, I will focus the discussion on a single case study - that of Turkish migrants in Europe and their relation to media cultures. The arguments will be very concrete and specific - rather than general and theoretical - but I hope that they will, as a consequence of their particularity, ground my argument. I am concerned, then, to work against the grain of the national imagination in order to think about how we might deal with cultural diversity more adequately - and respectfully. The point is to think beyond old stereotypes of 'the Turk', and to see what some Turks at least might have to teach Europeans about cultural diversity.

Transnational Turkish media – a case of imagined community?

In Europe now, it is possible for Turks to receive around twenty satellite channels, providing a great range of programming, from commercial and religious broadcasters, as well as that of the state broadcaster, TRT. The programming is extremely popular with the great majority of Turkish migrants. In Germany, this new Turkish viewing culture that has grown around satellite broadcasting succeeded in provoking considerable resentment and hostility. It has been said that the Turks have chosen to retreat into their own 'private media world'. And, consequently, that they are becoming 'dissociated from the social life of everyday [German] society' (Marenbach, 1995). The most extreme version of this argument has been elaborated in the alarmist writings of Wilhelm Heitmeyer. Here the discourse is centred on anxieties concerning cultural ghettoization, the dangers of 'Islamic fundamentalism', and the marking out of new 'lines of ethno-cultural confrontation and conflict' (Heitmeyer, in Heitmeyer et al., 1997, pp. 30–31). Heitmeyer's is an extreme expression of a panic that has been aroused through the creation of a new Turkish media space across Europe. But even among other, more liberal commentators, we suggest, it is possible to discern similar concerns and similarlygrounded responses. Even in their more liberal discourses, and albeit in a more tolerant way, what is at issue is still the maintenance of national cultural integrity and the cultural integration of minorities. In all the responses to transnational Turkish media, what seems to be at issue is how best to domesticate/associate/acculturate Turks who seem to be threatening to create their own separate cultural order.

There are all kinds of problematical assumptions in this defensive cultural-political agenda - this essentially national agenda. The most basic is the idea of the Turk as a problem in the European space (and, of course, this fits into a long European history of anxiety and fear about the difference that Turks bring with them to Europe). The problem that Heitmeyer and others have is really with their own imagined Turk – it is the problem of a cultural mythology of Turkishness that is for the most part their own projection. What they have created is the image of a homogeneous Turkish community inhabiting a unitary Turkish cultural space. And then they go on to make the further assumption that Turks in Europe must necessarily be 'taken over' by the new Turkish television channels they are watching – that they are passive and susceptible to the influence of their 'national' media system. The Turks are thought of as being inherently different from the Germans and other Europeans. And it is on account of this difference, it would seem, that they are vulnerable to Turkish media – and, through this vulnerability, disturbing and troublesome (*unheimlich)* for the Germans. In the case of the Turks, then, it seems that all kinds of old and discredited theories of media influence and effects are still suitable, and can still be strategically mobilized.

And, if it is asserted that Turks in Europe who watch satellite channels from Turkey do so because they want to be immersed in the culture of their 'homeland', it is not just the likes of Heitmeyer who think that this is the case. Certain elements in the Turkish media, too, seem to believe that this kind of service is what they are providing. This has certainly been a strong element in the thinking of TRT, the state broadcaster. And a recent marketing brochure promoting EuroD (*Reach 4.5 Million Turks in Europe*) makes the claim, with respect to European Turks, that 'European television leaves them where they are: Turkish television takes them home.' 'In Germany,' it is said,

> nearly a million televisions are turned [sic] into Turkish television by satellite during prime time every night. Turkish viewers overwhelmingly choose Turkish broadcasts whether not to forget Turkish or because they love Turkish pop-music or they find Turkish programming more meaningful. From politics to comedy, the Turks of Europe keep in touch with their roots by satellite, the only broadcast where you can really communicate with them. These viewers value Turkish television entertainment and they are willing and able to buy satellite dishes. In fact, survey results show that the loyalty of European Turks to Turkish broadcasting is unshakeable.

The value of Turkish media for Turkish audiences is, according to EuroD, to 'keep them in touch with their homeland.' Here too though of course, in a more affirmative sense than Heitmeyer's, the discourse centres on the value of national belonging. Here, too, it seems as if belonging to an imagined community – in this case the imagined community of the Turks – is the only basis on which it would be possible to make sense of viewers' engagement with the new transnational media culture

In the following discussion, I want to work against the grain of these nation-centric discourses. What I shall argue is that this image of programming 'from home' in fact works to obscure important complexities in culture and identity among the European Turks. We have to pose the question, then, of just what it is that Turks in Europe might actually be seeing, doing and experiencing when they are enjoying a 'full-fledged Turkish broadcast' on EuroD (or

Show TV or ATV, and even on TRT). Against the mythology of the Turks and their essentially problematical Turkishness, I want to propose a more sociological – and, let us say, a more mundane – approach. Not seeing Turks as a threat or as a problem, but simply as another group in the new European space – with a culture that is generally as ordinary as that of the Germans, the British, or the French. I consider what is currently happening to Turkishness in terms of a new cultural pluralization – we shall argue that there are actually many different kinds of Turkish culture and many different kinds of Turks (as there are of Germans, Britons and French citizens).

Transcultural sensibilities

It is significant that there should be such a belief that, when 'Turkish people' watch 'Turkish television', they are doing so as a unified community plugging into a unified cultural space. The fact that there is such a belief is indicative, I suggest, of how powerfully the homogenizing imagination of national community works to inhibit the perception of new developments and possibilities in culture and identity. But, however much Europe or Turkey imagines its Turkish migrants to be cohering around a homogeneous conception of Turkishness, behaving like a close-knit group, this proves to be far from what is actually taking place. I would maintain that, in the case of very many consumers of transnational television from Turkey, what they are experiencing is never straightforwardly and unproblematically the sense of being 'at home' or of 'keeping in touch with their roots'. And if we think carefully about what they are actually experiencing – how they are in fact using television – I believe that it then becomes difficult to sustain the rather simplistic notion that transnational television consumption could ever simply, and unproblematically, be about extending an 'imagined community'. Indeed, quite other cultural possibilities are there to be discerned – possibilities that are not at all to do with belonging, but, far more interestingly, we suggest, with cultural ambivalence.

At the present time there is a certain contestation, in which contrary forces are competing over the contemporary meaning of Turkish culture and identity; there is a contestation, that is to say, between a closed idea of Turkishness and more open and plural possibilities. Turks are involved in a process of testing whether they can continue to talk to one another now on the basis of their differences. And Turks in Europe are involved in an even more complex process of negotiation, having to position themselves in relation to both their changing sense of Turkishness and their experiences of the European societies in which they live. What I am going to argue is that this negotiation process involves Turks in *thinking* around issues of belonging, identity and culture. And, as thinking starts to problematize the supposedly given and fixed meanings of all these categories, we find that the possibility is opened up for them to become more aware of the always-provisional nature of cultural identity. They may become aware that the cultural community they imagined as being eternal and pre-given in fact turns out to be no more than an imagined – that is to say socially invented – community.

The recent emergence of the possibility of watching Turkish television in their own homes in migrancy actually puts Turks in a very different position from members of the host-society. And it is instructive to reflect on the differences of experience. For the host-community audience, the choice of channels and then of programmes is generally a matter of the confirmation of tastes. For the non-migrants, choices about television viewing generally take as their universe those channels and programmes which are in their own language, embedded in their own national references, and reflecting those practices, customs and habits with which they are familiar (and deciding to watch Eurosport, for instance, does not take them much beyond what they are already accustomed to). Watching television as a non-migrant, then, can be rather a sedentary affair, centred on a screen that reflects their sense of belonging – at-homeness – in an imagined community.

And, rather than reflecting on the different nature of the migrant experience of television viewing, the non-migrant mentality generally assumes that its own particular cultural practices are universal and normative. When its gaze is turned upon the media activities of the immigrant communities, it is simply and readily assumed that they are doing the same thing as the host-community audiences. And what underpins this assumption is the taken-for-granted idea of television functioning as a kind of cultural holding device. For the non-migrants, the expectation is that we all (must, should) live in singular cultural worlds (that is, imagined communities) – and that therefore a choice has to be made between, say, the German world and the Turkish world. The choice is always a stark one, between association and dissociation. Most observers of the consumption patterns of media among the migrant communities seem to work with this basic assumption. The primary aim is then to establish which of the holding environments the immigrants are opting for. Do they watch German TV? Or do they prefer Turkish TV? And if Turks are found to be watching 'too much' Turkish television, the non-migrant mentality automatically assumes that this must be due to their greater loyalty to their own tribe (due to their essential Turkishness) – and to a breakdown of the mechanisms intended to bind them into the German holding environment.

Even in the more sympathetic studies – ones that seek to exonerate the Turks from the charge of nationalistic withdrawal – we commonly find the same basic assumptions. Thus, in a study of the impact of cable and satellite television among Turks in Germany, Alec Hargreaves (1999, p. 11) finds that 'there is a broad tendency for the importance of Turkish channels to weaken through succeeding generations of the diaspora'. This weakening is regarded as being quite natural, in so far as 'second- and third-generation members of minority ethnic groups have generally come to speak the national language of the country in which they live far better than that of the country in which their migrant forbears have their origins' (ibid., pp. 10–11). The assumption is that, as new generations of migrants become progressively 'acculturated' into the host culture, then what they regard as their holding environment will inevitably switch, to become different from that of the parental generation – and so, therefore, will their patterns of viewing (they will watch more German television). Even in this more accommodating approach, then, what still seems to be operative is the perspective of integration, into one or other community – and what it is that is more accommodating is the benign assumption that, over an extended timescale, the forces promoting integration will naturally tend to prevail over those encouraging dissociation.

But this simple question of 'which television?' (Turkish or German?) is blind to what it is that is different, distinctive – and interesting – about the migrant cultural experience. In its concern only with the decisions that are made about which channel to turn on, which programme to watch, it overlooks the processes through which decisions come to be worked through. The assumption, in the case of the Turks, is that their responses to the new media environment come about automatically and unreflectively. It is as if the question of 'which?' is driven by a reflex to deny what is in fact a very active and complex process of engaging with television, and thereby to undermine the agency of the migrant viewers. In fact, when we examine the processes of decision-making among Turkish audiences regarding their television choices, we find that there is a constant process of negotiating taking place. In the case of these migrant viewers, their choices about which channels and which programmes to watch always involve them in difficult negotiations between spatially and historically different cultures and life-worlds. Their decisions about what they watch can involve them in often painful and tense processes of thinking about identity. Migrants are always compelled to be self-reflexive about the choices that they make.

Of course, there will always be habitual and unreflective aspects to their behaviour – tuning in to the Turkish stations as they once used to back in Turkey. But the experience of viewing then compels the viewer to always *think* about what is happening. The constant movement between cultural positionings – being distant from what is seen on the television, being part of another life experience in another cultural setting, and yet still being connected to Turkey – contributes to this reflection process. Consequently, identities no longer remain given and fixed – they are thought about, changed, abandoned, and re-claimed. What is different and distinctive about the migrant experience of television culture – and what would be instructive for cultural critics in the host societies, if they could only grasp this other kind of cultural experience – is precisely the experience of *thinking across cultural spaces*, with all the possibilities that this then opens up for thinking beyond the small world of imagined communities.

How does this kind of mobile thinking manifest itself? In the research that I have undertaken in London (with Dr Asu Aksoy) (see, for example, Aksoy and Robins, 2003a, 2003b) – in both focus groups and individual interviews – what comes across most strongly is the way in which the participants reflect on their fantasies, re-experience their frustrations, articulate the tensions between their spontaneous and deliberative responses, and work through their conflicting thoughts about their difficult relationship to Turkish (television) culture. One of the ways in which migrants think about their relation to Turkish culture is through the modality of fantasy. They may choose to seek out those kinds of programmes that convey an ideal image of Turkey and of Turkishness. 'I love old Turkish films on television,' said a man in his late thirties, 'they take me back to Turkey' (Focus group, London, 3 March 2000). 'I sometimes ask myself,' said another man,

> What does a person outside his country miss the most? We go out for picnics occasionally – if you can call it a picnic, since we can't find trees to sit under and we can't start our grill. It is at these moments that I start thinking about our meadows back at home, our pine trees, our water. I'm from the Black Sea region and I remember our

> cool water falls, our sea. I wish my children could see the plateaux, the summer feasts, and learn about our customs and traditions...I wish these were on Turkish television (Focus group, London, 29 April 1999).

Again and again, people express their heartfelt desire to see a Turkey – whether it is in nature, childhood memories, or the 'old days' – that would make them happy. They look to television as the most obvious place to reflect back to them their ideal sense of Turkey – a Turkey that is longed for, desired, missed.

Sometimes fragments of this desired image of the ideal Turkey can be picked up from old movies, or from programmes that celebrate special events that have shaped Turkish consciousness. What seems to appeal is a sense of purity and unity – hints of a world before degeneration had set in. A young woman, who came to Britain when she was seven, told us how these kinds of programmes could carry her into 'a rose-tinted world'. It is possible, she said, to 'to become lost in dreams, imaginings...it gives you a very sweet sense' (Focus group, London, 3 November 1999). A middle-aged woman from the Black Sea region, who has been in London for a decade, is happy to watch even the Islamic-oriented Kanal 7 – and to let go of what she regards as her secularist principals – just so that she can hear Turkish folk music: 'When I listen to the regional tunes I am absorbed, lost in my old days. I wish the other channels would reflect our culture. Then everybody would be happy...I sit for hours watching, absorbed in my dreams' (Focus group, London, 10 February 2000). 'We expect Turkish television to reflect those life styles that belong to our childhood...' says another focus group participant (Focus group, London, 10 February 2000). So television programmes are scrutinized with this needy eye. There is this tremendous need for objects on which to project and preserve an ideal image of Turkey.

These idealizations are a significant aspect of the migrant experience. And sometimes, it is clear, such imaginings can be overpowering. But what we have found in talking to Turkish migrants (particularly in focus group settings) is that, as they articulate such idealizations, they are commonly doing so in a self-aware and self-reflecting way. They are conscious of their need to elaborate compensatory mechanisms. And they know very well that their idealizations are rooted in a past that has gone; a past that can never be brought back to life. Even when they are dreaming in this way, then, our informants are *thinking*. They are able to also stand at a distance from their dreaming selves. They are thinking about the significance of this dreaming experience. They are thinking about the aspect of themselves that is happy to be engaged with the world of phantom identities.

And they are also thinking, at the same time, about how these idealizations and fantasies relate to the actual reality of Turkey. For they generally recognize that the ideal that they sometimes invoke does not at all correspond to the Turkey that they know from direct experience. And this recognition is always a source of disappointment. 'After three years being away, when I went back to Turkey I realised that it was definitely not the place that I left behind. I saw that it was not the place that I carried in my head, in my brain...[I]t is not a place that I would search for, where

I would desire to be...' (Focus group, London, 10 February 2000). It is important here to stress that it is not only through television and the media in general that Turkish people get a sense of Turkey as it is now. As communications and transportation become cheaper, people phone people they know in Turkey more, they visit their relatives, or just go to Turkey for a holiday. They have a direct sense of it. And, paradoxically, having a direct sense of Turkey can make these Turks more detached. 'I feel like a foreigner in Turkey,' says one young woman, who came to Britain ten years ago, at the age of seventeen, 'I can't recognize the money, I find shopping very different – in other words I find everything different there. For this reason I'm glad to be here [in London]. I feel I've grown up here. I wouldn't think of going back. I would go for visits, but not for good' (Focus group, London, 18 May 1999). First-generation migrants, as well as younger ones, carry this fear – that, if they returned to Turkey, they would no longer fit in, they would not be able to absorb and adapt to the changes that have taken place in the country since they left. As well as having a fantasy dimension, then, the thinking of migrants about their relationship to Turkey and to Turkishness is also informed by a strong reality principle. They are able to move across modalities of experiencing and of thinking about experiences.

If television allows these viewers to grasp compensatory images, and to think about their experiences by means of these images, there is a great deal more than this happening. There are quite other kinds of relationship with the complicated world of the screen, too. What Turks in Europe may come to realize is that the very thing that is desired, to bridge the emotional distance from Turkish everyday life, can also at the same time involve experiences of distantiation. Rather than being able to identify with a comforting Turkish community, they are in fact likely to be driven to feelings of ambivalence, discomfort and frustration. News programmes, which seem most of all to be about connecting with Turkish actuality – it always seems the most necessary kind of programming – prove to be particularly distressing. 'As you sit down to eat your food with your family,' says one informant, 'you are watching television, and you see a young girl being carried, covered in blood. The news brings demoralising stories' (Focus group participant, London, 18 May 1999). 'It is intolerable', says another. 'We don't like Turkish television between five and seven [in the evening]...because that is when the news is on' (Focus group, London, 3 November 1999). And again, 'there's too much violence on Turkish television, there's nothing good about watching the Turkish channels, only the stress' (Focus group, London, 10 February 2000). To watch the news can be to realize things that previously seemed impossible about one's country. For many Turks, then, to become synchronized with Turkish realities can be to put themselves in a very unsettling position – to become removed, and even alienated, from the thing that they thought was theirs.

Watching Turkish television seems to gratify the desire to 'be there', to be connected with everyday Turkish rhythms and realities. Before the advent of Turkish satellite and cable services, Turks in Europe could only relate to the Turkey that they had left behind them. As one of Kanal D's former news coordinator, pointed out, 'Turks were getting disconnected from Turkey, and they were stuck in a frozen history. As if the clock had stopped, they were repeating what they had when they left Turkey.' 'But now,' he adds, 'it is not like that. They are following the Turkish agenda day by day' (Interview, Istanbul, 29 December 1998). Now it seems possible for the

Turks in Europe to be at the same time in the Turkish space. In one respect, then, the appeal of Turkish television is, indeed, about being synchronized with Turkish time and events. But, at the same time, we would argue, Turkish viewers are constantly made aware that their engagement with this Turkish cultural space is still different – they are continually reminded that it can only ever be 'as if' they are there. The sense of synchronization with the time and events of Turkey is always a disrupted and disturbed sense – always in some way an alienated sense.

The complexity expressed by many of our informants comes in part from their condition of involvement in two cultures. But, their insights also come, we believe, as a consequence of what they also experience as their sense of distance from both cultures. What are crucial are precisely the possibilities afforded them by cultural distance (which is the antithesis of the stultification afforded by the cultural intimacy of imagined community). They have a certain freedom to think because they are not at home in two spaces at once. It is actually their experience – their double experience – of distance and detachment that enables them to think between spaces. It is important to think what this detachment is all about. The imam of a Turkish mosque in London expresses the experience strikingly:

> When I first came here in 1994, I was temporary. I didn't bring the children and I missed Turkey very much. Every evening I used to sit in front of the Turkish television and start with the first news of the evening, whichever channel it was on, and then move to the next one on a different channel, and then to the next one, in succession. Every evening I did this until the late hours, watching the same news that I had heard earlier, again and again. You know what there is on the news: people dying of hunger, people committing suicide, people throwing themselves out of their windows because of their debt, and so forth. We used to see the same things back in Turkey on the television, but probably not with the same kind of serious and devoted interest. We probably did not pay as much attention there to the news. I phoned a friend of mine in Turkey and asked him how he was managing to survive there. 'Is life continuing there?' I asked. 'What's the matter,' he said, 'why are you asking me this?' I replied, saying that as I watched television here I had the impression that life there was intolerable. Can you believe it, in three months I built up this impression of life in Turkey by watching Turkish television. Only three months before I was someone living there, and in such a short time I came to be asking about the life conditions there. I think that the image of Turkey on television and its reality are very different. Everybody in Turkey in fact goes about their life and their business in a normal way (Focus group meeting, London, 29 April 1999).

What this informant comes to recognize is that Turks in Europe do not actually watch Turkish television as they would if they were in Turkey. As the former Kanal D news coordinator put it, 'when you are in Turkey you know that what is being reported on the news is rather exceptional, and you have a sense of reality due to the fact that you live in the same environment. When you are abroad your sense of the difference between what is being reported and what constitutes reality is collapsed. So, you begin to equate them' (Interview, 29 December 1998). Turks discover the possibility of alienation from a deeply cherished image of Turkey. But, at the same time, however, they are able to become more lucid about the nature of cultural identities, and particularly the identities associated with belonging to imagined communities. Through their cultural distance, then, it is possible for Turkish viewers in Europe to become aware of the constructed nature of Turkish realities – they find themselves in a position to become deconstructionist critics of the media culture they are consuming.

Cultural diversity and public culture in the new Europe

In this discussion, I have taken a small, focused case study in order to explore a central issue concerning cultural diversity and public culture in contemporary Europe – a small case study as a way in to discussing what is a very big agenda. What I now want to do is to extrapolate some of the broader and more general points that seem to emerge from this account of Turkish satellite television. What, for example, could we learn from the German responses to changing viewing habits among Turks? And what might we learn from the banal, everyday use of transnational television by Turks in Europe – the ways in which they have fitted transnational television into their cultural lives?

The first point – the core issue addressed in this chapter – concerns how powerful the national imagination (the paradigm of imagined community) is when it comes to thinking about audio-visual spaces; it is apparent in both the attitude of Heitmeyer and that of certain Turkish media interests. What we have to address is the taken-for-grantedness of the national frame. Many media researchers have drawn attention to great expansion of transnational broadcasting across Europe (Chalaby, 2002), with increasing media transgressions of national audio-visual boundaries. These developments have been increasingly undermining the sovereignty of national media systems. Thus, Price (2002, pp. 4–5) writes of 'a vast remapping of the relationship of the state to images, messages and information within its boundaries,' and emphasizes the need for new thinking about these developments 'in the face of the inevitable persistence of older modes of description and analysis' (that is, the agenda of imagined community still persists). What are coming into existence are complex new audio-visual geographies (Sinclair, Jacka and Cunningham, 1996). And yet, notwithstanding these complex transformations, we find constant re-assertions of the centrality of the national paradigm. Thus, in their critical engagement with media and globalization, Waisbord and Morris reaffirm the value of nation states in the face of contemporary change. 'Collective identity,' they maintain, 'is still fundamentally tied to the state as both a power container and an identity container' (2001, p. xv). Similarly, in a defence of public service broadcasting, Curran (1998, pp. 193–4) defends the virtues of imagined community. He continues to validate public service because 'it upholds the values of community', promoting a 'culture of social integration and mutuality'. What is clear in many contributions is that it is difficult to address the challenges of transnationalization other than through the old paradigms of the national order.

The national paradigm has a clear and potent appeal. It has been the prevailing model for audio-visual cultures for so long that it has come to seem the 'natural' model, and the only frame developing media regulation and policy. The images of 'container' and 'community', as in the above quotations, have a certain resonance with their evocations of security and familiarity in a culture held in common. But it is precisely this cosiness that is now problematical, I want to suggest. In the context of a rapidly changing, transnational Europe, this container imagination can no longer be regarded as entirely functional. The fundamental problem with this paradigm is that it conceives a cultural map in which different cultures are both internally homogeneous and bounded from other (external) cultures. So, as Curran (1998, p. 191) puts it, 'broadcasting stages a collective conversation about common social processes...society communes with itself'. It is 'organised in a way that both reflects and fosters a culture of mutuality and responsibility to others' (1998, p. 193). And it is about commemorat[ing] the shared life of a community [and]... affirm[ing] its sense of continuity' (1998. p. 194). The problem is that the conversation is only between members of a relatively homogeneous community; the mutuality is (primarily) to its members, the insiders; and the sense of continuity is its alone. What should be recognized is that this particular model of culture – in Curran's account we have a more benign version of this cultural paradigm, whilst Heitmeyer demonstrates a more militant variant – has no place for heterogeneity and complexity in its philosophy and spirit. It cannot easily acknowledge and deal with diversity (other than through the strategy of cultural minoritization). The paradigm was sustainable – with all its inadequacies – as long as the national paradigm remained unquestioned. But now, when times are throwing up much more complex forms of cultural experience, there is a need for more open and inventive ways of responding to change. A European approach to diversity in the media and cultural sectors must involve an approach that sees diversity not as a problem, but a resource and opportunity, an approach that can accept diversity – rather than consensus and confirmation – as being at the heart of its project and imagination.

The second point that I want to make here concerns the way forward in thinking about cultural diversity at a transnational and transcultural level. What I wanted to bring out in the case study of Turkish migrant usage of satellite television was something about the possibilities of transcultural imagination. The development of so-called diasporic media (MBC, Televisa, Asianet, TRT INT, TV Ahmadiyya, Al Jezeera Television, or the Turkish channels I have referred to) – all of them broadcasting to particular cultural and language groups across national boundaries – provide a good focus for considering transnational diversity. For it seems that particular ethnic, religious, and language-based cultures are finding in transnational broadcasting the kinds of services that they have been missing in national broadcasting schedules. Through these transnational developments, established models of public service broadcasting and of minority programming are potentially being subverted – models that have been based on the national organization of media policy and regulation. The new reality concerns the re-positioning of minoritized ethnic and cultural groups in the context of new transnational and diasporic audiences.

In my discussion of Turkish migrants I wanted to suggest that we might learn something about this new transnational sensibility. My aim was to bring out something of what is happening as people learn to think across different media and cultural spaces. The changes in demand seem to reflect a new kind of identity position among viewers. In part this is a consequence of the ability now, as a consequence of cheap airfares, and the relative proximity of Turkey to London, to travel to and fro. But it is also present at the mundane, experiential level of television viewing. People feel themselves to be part of both Turkish and British cultural spaces. They want to operate across both spaces, with no singular allegiance to one or the other. These audiences have begun to develop new practices, then, that involve them in an imaginative and intellectual mobility across cultural spaces. We need to take heed of this new cultural positioning, in which audience members release themselves from exclusive membership to a single 'imagined community' (I say this because most research into so-called diasporic media insists that it is all about the pleasures of imagined community and long-distance nationalism). There is something new in this sensibility and cultural orientation. It is surely a phenomenon that should be of interest to media policymakers, and also to those concerned with broader cultural policy in the new Europe.

Cultural diversity and media policy

What is called for is a new political and cultural geography for media policy and regulation. Important new issues are being opened up concerning cultural rights, cultural diversity and public culture, on a basis that now exceeds the national framework:

There are important **cultural** issues concerning 'cultural rights' and multiculturalism. What are the cultural implications when sizeable migrant communities cease to watch the national channels of their 'host' country for cultural diversity strategies and policies? Does the concept of 'minority' programming cease to be adequate for addressing audiences that have until now been categorized in this way? How should multiculturalist policies in broadcasting be re-invented in the age of transnational broadcasting?

And, of course, there are also very significant **political** questions to be raised concerning the public sphere and the future of public service philosophy. How can these new transnational services be incorporated into models of the public broadcasting sphere? What is the significance of these new media for public-service ideals, nationally but also increasingly at the European scale? How should the public broadcasters be responding to the increasing penetration of transnational broadcasters into the mainstream audiences? What are the scales of intervention for media policy agencies given the transnationalization process?

The processes of media transnationalism are posing a whole new set of questions with respect to cultural diversity issues and the emergence of new public spheres/cultures in the European space.

But, if developments in transnational broadcasting are thus raising new issues about both media regulation and cultural policy, we may say that there is no constituency for discussing what the implications of this new development are. There are now possibilities for the institution of what could be an interestingly and productively new transnational European map. And yet media policy remains predominantly national, and has not begun to consider the implications of national citizens watching a complex new array of programming. On the other hand, those concerned with cultural diversity and multicultural issues have not been sufficiently engaged with developments in the media sector. The issues raised by the transnationalization of broadcasting may be said to be falling between stools. There has not yet emerged an agenda – and an imagination – to deal with the challenges for Europe of other-than-national dynamics in media industries and cultures.

References

Aksoy, Asu and Robins, Kevin (2003a) 'Banal transnationalism: the difference that television makes', pp. 89–104 in Karim H. Karim (ed), *The Media of Diaspora: Mapping the Global*, London: Routledge.

Aksoy, Asu and Robins, Kevin (2003a), 'The enlargement of meaning: social demand in a transnational context', *Gazette: The International Journal for Communication Studies*, 65(4–5): 365–388.

Cardiff, D. and Scannell, P. (1987) 'Broadcasting and national unity', in J. Curran, A. Smith and P. Wingate (eds), *Impacts and Influences: Essays on Media and Power in the Twentieth Century*, London, Methuen.

Chalaby, J. (2002) 'Transnational television in Europe: the role of pan-European channels', *European Journal of Communication*, 17(2): 183–203.

Hargreaves, Alec (1999) 'Transnational broadcasting audiences: new diasporas for old?' Paper presented to the Workshop on Media in Multilingual and Multicultural Settings, Klagenfurt, 11–13 November.

Heitmeyer, Wilhelm, Schröder, Helmut and Müller, Joachim (1997) 'Desintegration und islamischer Fundamentalismus: Über Lebenssituation, Alltagserfahrungen und ihre Verarbeitungsformen bei türkischen Jugendlichen in Deutschland', *Aus Politik und Zeitgeschichte*, 7–8: 17–31.

Marenbach, I. (1995) 'Die Welt aus türkischer Sicht', *Süddeutsche Zeitung*, 19 May.

Price, M. E. (2002) *Media and Sovereignty: The Global Information Revolution and Its Challenge to State Power*, Cambridge, Mass.: MIT Press.

Scannell, P. (1989) 'Public service broadcasting and modern public life', *Media, Culture and Society*, 11(2): 135–166.

Sinclair, J., Jacka, E. and Cunningham, S. (eds) (1996) *New Patterns in Global Television: Peripheral Vision*, Oxford: Oxford University Press.

Waisbord, S. and Morris, N. (2001) 'Introduction: rethinking media globalisation and state power', in N.Morris and S. Waisbord (eds), *Media and Globalization: Why the State Matters*, Lanham, Rowman & Littlefield.

Meanings of Money: The Euro as a Sign of Value and of Cultural Identity

Johan Fornäs

Coins and banknotes are not only economic signs of value but also symbolic signs of cultural identity of geopolitical unities. They function as communicative forms and minimalist mass media. Though carefully planned in terms of design, the euro money introduced in January 2002 offer an embryonic premonition of a vision of European identity, disseminated from official institutions based in the state and market systems, but reaching deep into the wallets of daily life. Critical attention to symbolic identifications encoded in everyday artefacts may contribute to uncovering key late-modern dream images and to highlighting ideological forms that normally pass unnoticed, thereby de-naturalizing what Michael Billig has called 'banal nationalism' (1995).

The institution of money is an organizing tool for the circulation of goods and services and for binding society together. It is a means of communication intended for the combined use as unit of account, means of payment and store of value. In order to fulfil these economic functions, currency must contain texts, images and patterns that make them interpretable as money, ensuring authenticity and representing value, nationality and time of issue. The ways in which to display and safely guarantee their value can be varied and elaborated in response to a wish to make them more visually appealing, or to add other layers of meaning that reflect how economic values and the country of origin are understood by its monetary authorities and ordinary citizens. Money designs are at once aesthetic and economic, carriers of meaning as well as of financial value. They signify not only 'frozen desire', but also forms of identification. Though regulated by political-economic systems, there is a surplus of meaning, in that these extra-economic, symbolic use-values of money open up a space for imagination and interpretation uncovering unintended and unconscious layers of identifying meaning. Even though money is not *primarily* a communication medium in the same sense as television, books or CDs, it is fully possible to study *the meaning of money as media*.

Money artefacts mediate in a double sense, their economic interconnecting of society reinforced by their role as symbolic identifiers.[1] But there are surprisingly few studies of this latter cultural aspect, and a striking gap between numismatic studies and media/cultural studies. In *A Flutter of Banknotes* (2001), Brion and Moreau survey the motif history of European paper money. Notes have often showed antique gods or predominantly female allegorical figures representing human virtues or aspects of activity related to the idea of progress: commerce, industry, agriculture, science and art. Symbols of permanence or vigilance were meant to inspire confidence: anchors, hives, towers, open eyes, lamps or cocks. Other banknotes depicted national symbols: coats of arms, heraldic beasts, portraits of monarchs, or more indirectly motifs relating to folklore, local landscapes or place-bound mythology. Portraits in a realist style have dominated since the Second World War, and national figureheads from art, philosophy and science became prominent features from the 1960s. In general, banknotes tend to reflect the main values of the issuing societies: 'faith in progress, the virtue of work, social harmony, the greatness of a nation', offering an insight into 'the great founding myths of Western society' (Brion & Moreau, 2001). Today, the euro is a particularly fascinating site where basic European identities are represented but also made.

Another study, by Jacques E.C. Hymans (2004), investigates currency iconography as indicator of collective identities in Europe since the early nineteenth century, using a database of 1,368 notes from all the fifteen member states. Its main finding is that time (period) appears more decisive than space (nationality) for paper money images, indicating that states express a transnational spirit of the times rather than unique national identities. Inspired by Ronald Inglehart's theories of cultural shifts, Hymans discerns that in these fifteen countries an overall trend for the social focus to move from state over society to the individual, and of basic norms to move from tradition over material goods to post-materialist values. He sees the paper euro as confirming these trends, but in this case the focus on banknotes precludes the national differences that may only appear on coins. It is also, in practice, often difficult to decide whether a specific symbolic motif should be understood as a state, societal or individual actor, or reflecting traditional, materialist or post-materialist values. For instance, both antique myths and classical artists connect to traditions but may still be interpreted in contemporary terms, and a scientist can embody both materialist and post-materialist life goals. Still, these studies offer a useful historical background to today's euro iconography.

Constituting Europe

European integration takes place on several arenas. The 2003 draft of a treaty establishing a constitution for Europe expresses a conviction that 'while remaining proud of their own national identities and history, the peoples of Europe are determined to transcend their ancient divisions and, united ever more closely, to forge a common destiny'. It finds the European Union 'reflecting the will of the citizens and States of Europe to build a common future' and based on 'the values of respect for human dignity, liberty, democracy, equality, the rule of law and respect for human rights'. It offers its citizens 'an area of freedom, security and justice without internal frontiers, and a single market where competition is free and undistorted', while promising to 'respect its rich cultural and linguistic diversity', and ensuring that 'Europe's cultural heritage is safeguarded and enhanced'. Five 'symbols of the Union' are specified:

- The flag of the Union shall be a circle of twelve golden stars on a blue background.
- The anthem of the Union shall be based on the *Ode to Joy* from the *Ninth Symphony* by Ludwig van Beethoven.
- The motto of the Union shall be: United in diversity.
- The currency of the Union shall be the euro.
- 9 May shall be celebrated throughout the Union as Europe day.[2]

The European *flag* goes back to the Council of Europe in 1955, and was adopted by the EU institutions in 1986. Its gold star circle is supposed to symbolize 'Europe's unity and identity' and the 'solidarity and harmony between the peoples of Europe'. The number of stars has nothing to do with the number of Member States, but was chosen as a traditional 'symbol of perfection, completeness and unity'. Twelve is the number of months in a year and the number of hours shown on a clock face, thus connoting the dynamism of time, and the circle is often used as a symbol of unity.

The European *anthem* is the famous theme from the final movement of the 1823 symphony where Beethoven used Schiller's idealistic 1785 vision of the human race united in brotherhood. It was adopted by the Council of Europe in 1972 and became the official EU anthem in 1985. 'Without words, in the universal language of music, this anthem expresses the ideals of freedom, peace and solidarity for which Europe stands', says the EU website.

Europe *day* is 9 May. This was the date in 1950 of the 'Schuman declaration', where the French Foreign Minister Robert Schuman proposed the creation of an organized Europe to abolish the divisions that had caused such disastrous wars. France and Germany should head the other European countries in pooling together their coal and steel production as 'the first concrete foundation of a European federation'.[3] In 1985, the Milan Summit of EU decided to celebrate 9 May as Europe Day, emphasizing values of peace and solidarity that 'find expression through economic and social development embracing environmental and regional dimensions which are the guarantees of a decent standard of living for all citizens'. Europe Day is to be used for 'activities and festivities that bring Europe closer to its citizens and peoples of the Union closer to one another', helping European citizens to 'feel at ease in the "European home"'.

The European *motto* 'united in diversity' (at first: 'unity in difference') was accepted by the European Parliament in 2000, selected from proposals sent to a website by some 80,000 school pupils. It had previously been the motto of the European Bureau for the Lesser Used Languages, and is a key theme in today's EU discourse.

Together with the euro, these symbols, jointly identifying the political, economic and cultural entity of the European Union, are integrated in a standard stock of national symbols, and combine to work on several levels: visual, aural, verbal, economic and temporal. A sixth symbol might well be added, namely the EU passport, signifying 'European citizenship'. While the flag, anthem, day and motto have a more limited and purely symbolic or discursive use, the symbols for money and citizenship each have a double function, as both symbolic expression of identity and material tool of integration – in one case economic, in the other political.[4]

The modern European project has emerged as a learning-process from the violent experience of deep internal differences. Many, including Jürgen Habermas (2001), have emphasized the need for economic and political EU institutions to be anchored in a combination of mediating, civil society-based 'sphere of publics' and collective European identifications. These are emerging only slowly, and a set of five symbols is but a very minor step. Still, they do mean something. European identifications emerge in everyday interactions among people, but are supported by specific public channels and symbols afforded official status. Each such symbol may in itself appear trifling, but in combination and context, they provide a guiding image of what Europe is or may become.

Experiences and imaginations of the character and role of Europe in the world are formulated in literature, art, songs and films, on television and the Internet, but also in the faces of the euro. It is interesting to note that this currency is explicitly treated by the EU constitutional treaty as an identifying symbol, and not only as an economic tool. On one hand, this testifies to a 'commercialization' of the European project: an explicit acknowledging of the central role of the market economy in the union, not only as a hidden linking mechanism but also as a cherished, quasi-sacred item. On the other hand, it also simultaneously expresses a 'culturalization' of the economy, acknowledging the fact that even money as aesthetic material objects are part of an experience industry. Such money symbols are even more omnipresent than the other four symbols, and therefore deserve closer scrutiny. They belong to a kind of 'unflagged' or *'banal' supranationalism*.[5]

Introducing the euro

On 1 January 2002, the seven different values of euro banknotes and eight values of coins were introduced in twelve Member States of the European Union, to be used by almost 300 million Europeans. Monaco, San Marino and Vatican City participate in the euro currency with their own coin designs, through a special agreement. In all, 50 billion coins and 14.5 billion banknotes were released, with a total value of over €664 billion. The name 'euro' was adopted in 1995 as a successor of the previous European currency unit 'ecu' which for Germans sounded like 'ein Kuh' (a cow) and thus was deemed to invite jokes. The € symbol is based on the Greek epsilon, referring to the origins of European civilization, with two horizontal bars symbolizing intended monetary stability. The banknotes look the same throughout the EMU area, while the coins have the front side (obverse) common to all twelve countries and a rear side (reverse) specific to each country (for images of all euro banknotes and coins, see http://www.ecb.int/bc).[6]

In 1995 the European Monetary Institute (EMI), forerunner of the European Central Bank (ECB), selected two themes for the euro *banknotes*: 'Ages and styles of Europe' and a broader theme of 'abstract/modern design'. For the first theme, the features to be depicted on each of the seven denominations should represent a specific period of European cultural history. The whole process was a combination of EMI/ECB decisions, open competitions, qualitative interviews with European cash handlers and ordinary citizens, and a jury of experts in marketing, design and art history. They used criteria of 'creativity, aesthetics, style, functionality, likely public perception and acceptability (in particular the avoidance of any national bias and the achievement of a proper balance between the number of men and the number of women portrayed on the banknotes)'. In 1997, the final designs were created by graphic designer Robert Kalina of the Austrian National Bank. Apart from basic information such as the value and the name of the currency in the Latin and Greek alphabet, they include a value-specific combination of the twelve EU stars with a set of windows and gateways from seven architectural periods: Classical (5€), Romanesque (10€), Gothic (20€), Renaissance (50€), Baroque and Rococo (100€), Iron and Glass style (200€) and Modern twentieth century architecture (500€). These architectural elements are deliberately abstractly designed in order not to signify any particular building from any specific country, but to synthesize features that unite the whole continent. Excluding human figures expediently solved the problem of gender and ethnic balance. The windows and gateways symbolize 'the European spirit of openness and cooperation', while the twelve stars represent 'the dynamism and harmony between European nations'. The reverse of each banknote features an unspecific bridge from the corresponding period, symbolizing 'the close cooperation and communication between Europe and the rest of the world'. There is also a map of Europe, with tiny dots for the extra-European colonial territories of France, Portugal and Spain that also use the euro.

As for the *coins*, their eight different obverse sides have a motif created by. Luc Luycx of the Royal Belgian Mint, who won a European-wide competition. They depict the value, the name 'EURO' and different variants of the EU map and twelve stars linked by parallel lines. The one, two and five cent coins supposedly show 'Europe's place in the world', by having a map of the entire globe with Europe in the centre. The ten, twenty and 50 cent coins depict 'Europe as a group of individual nations' by showing each country as a separate island. 'A united Europe without frontiers' is represented on the 1 and 2 euro coins, with a common EU map. These three variants are also differentiated in colours and general design, so that the coin series consists of three value groups. These three variants together tell a narrative starting with entering Europe from afar, noting its place in a global context, then focusing its internal diversity, and finally watching it unite into a coherent entity. The lines between stars imply a kind of unique and holy 'star quality' for each state with an emphasis on the linking work of their union. The two-sided coin series is declared to express the motto 'united in diversity', the common obverse side symbolizing the unity of the European Commission, whereas the national reverse sides represent the diversity of the European Parliament. It is notable that the obverse sides symbolize pure financial value in the form of simply a number, whereas more elaborate aspects of cultural identity are placed on the reverse sides. All coins may be used in all EU countries, resulting in a circulation of national signs between the states as well as to all other countries where the euro is used. Each national coin set tends to dominate the circulation of money in its respective country, but through trade, travel and tourism, national circuits leak into each other, so that citizens will, from time to time in daily life, also encounter images from elsewhere. International contacts thus leave traces in a varying co-presence of national coin variants. The co-presence of a range of national symbols in any single EU citizen's wallet is a reminder of the coexistence within the boundaries of this union of previously alien and exotic regions.[7]

United by diversity

As one of the five official symbols of the union, the euro as such is a main unifying element, stressing the 'unity' in the European motto 'united in diversity'. The euro is formally a coherently designed currency with strong unifying elements, including all banknotes, coin obverses and the general frames for the national reverses. By deciding values and sizes, and thereby forcing the national reverses into specific preferred patterns, the coins' common sides work as a common EU denominator. However, scrutinizing the contents of the national messages on the reverse sides of the coins discloses a more divergent picture.[8] Is the main message the intended one of unity in difference – or rather a candid agenda of difference in unity? Which kinds of unity and which kinds of diversity are actually acknowledged? Is abstract universalism, shared European values, international cooperation, transnational (or even postnational) relations, or distinct national communities most prominent among the coin reverse motifs?

Differences

Some general patterns may be discerned on these national reverses, concerning (a) currency values, (b) money genres, (c) country groups and (d) historical changes. As for the first aspect, there is no uniformity in how the countries have divided the coins into sub-series. The common obverse designs tend to favour a 3+3+2 tripartition, paralleled on the reverse sides of France, Germany, Luxembourg, Portugal and Spain, while all other countries make other subdivisions. Different countries also construct very different 'money-stories' from lower to higher values, based on contrasting hierarchies. One may discern a dominant story from a basis in nature and technology up to culture, myth and ideas on the highest values, reflecting a possibly typical European dualist hierarchy of body/soul or base and superstructure, which has both materialist and idealist versions, depending on whether the low material basis is seen as foundational or subordinated.

Genres of motifs

A handful of main money genres may be distinguished:

1. *Rulers* are shown by Belgium, Luxembourg, the Netherlands and Spain. All the monarchies – and only these – display their rulers, leaning towards an ancient tradition of authorizing money values by showing the ruling head of a clan, empire or nation. In modern

republics that practice has become less useful, due to a combination of recurrent shifts of power and perhaps also to some small degree an egalitarian spirit of democracy that shuns displaying such clear symbols of state power as national symbols. Hereditary monarchies have stabile heads of state, at the cost of stripping these anachronistic institutions of all essential instruments of real political power. It is slightly paradoxical that these monarchs nowadays have almost no political power, being reduced to purely symbolic signs for their nation states. But precisely this makes them doubly useful as money motifs, and perhaps the most easily accessible and in a way uncontroversial choice. They fuse aristocratic historical roots with late modern entertainment business and popular culture. If the specific monarch depicted has had no personal role as transnational bridge-builder, this motif is bound to the old European system of nation states out of which inter-national systems like the UN and the EU were once born, but else contains no other, more innovative or up-to-date transnational impetus.

2. *National symbols* are selected by Finland, France, Germany, Ireland and Portugal. Heraldic animals, coats of arms and other traditional symbols that have been monopolized by certain states fill similar functions as the rulers' faces, and are equally old as money symbols. While escaping the anthropomorphization of power, they reproduce feudal legitimation narratives of nation building. All national symbols to some extent have transnational roots and routes: they have travelled and branched off in various directions, and are never undisputedly local. Benedict Anderson (1991) has pointed out that, being much older than nation states, churches as well as royal houses remain particularly promiscuous in that respect, even when the latter are subsumed under national authorities and bound to their names – the King of X is often closely related to the Queen of Y. For an Irish citizen, the Celtic harp probably is a univocal and deep-rooted image for the Irish nation, but in Wales, Brittany or Galicia it might well intersect with other local traditions, in Jewish tradition it rather recalls King David of the Old Testament, and for a Greek or a Finn who gets such a coin in her purse, it may well be understood as just a nice old instrument that shows the universal reach of music. The question is about which kinds of such lines are drawn through the choice of such symbols: political, military, commercial, cultural, etc. Whereas the harp implies harmony and communication, crowns and seals signify power and authority, eagles and lions, like coats-of-arms, connote violence and military force, and plants have naturalizing meaning-effects of growth, care, boundedness to the soil, etc. Some national symbols have universalistic overtones. In the French imagination, 'liberté, égalite, fraternité' is a universal motto, spread through colonialism and anticolonial republicanism, right up to the UN declaration of human rights.
3. *History*, in the form of cultural or political artefacts and individuals, appears on coins from Austria, Germany, Greece, Italy and Spain. There are many subtypes in this category, as history contains many things with highly divergent implications. Political events, leaders or buildings may relate to key moments of nation-formation, in which case the signification of such motifs come close to the previous ones. Social, scientific or aesthetic heroes or works have a more crossover status, tending to move across borders and have global importance. A range of other differences appears depending on whether persons, events, buildings or other kinds of artefacts are depicted. Buildings are more fixed to a place than paintings or

people who can travel across boundaries, but they may on the other hand easily be visited by many and become widely known and loved, not least through modern mass tourism. Historical motifs tend to be selected to represent various regions within the nation, ages of national splendour and kinds of achievement, so that they, taken as a whole, represent the moral, intellectual and cultural strength of a country. Taken as a whole, Europe shows two political freedom fighters (the Greeks Capodistrias and Venizelos) and one peace activist (the Austrian von Suttner) who also is the sole woman honoured in this way by the EU, three literary authors (the Greek Velestinlis-Fereos, Italian Dante and Spanish Cervantes) and one composer (Austria's Mozart). This slight dominance for the cultural domain is increased when buildings are added, with three mainly political (the Austrian Belvedere Palace, the German Brandenburger Tor, the Italian Castel del Monte) against five cultural – mostly religious – ones (the Austrian St. Stephen's Cathedral and Wiener Secession building, the Italian Mole Antonellina and Colosseum, and the Spanish Santiago de Compostela cathedral). Greece's three ships express economy and trade but also transports of other kinds, military as well as civil. Adding other human artefacts further emphasizes the cultural face of Europe, with Italy's wide range of monuments and artistic works (Botticelli's Birth of Venus, Boccioni's futurist movement forms, Marcus Aurelius equestrian statue, Da Vinci's Vitruvian man, Raphael's Dante). It should also be noted that the Spanish Santiago da Compostela cathedral as well as Cervantes are of course as much rooted in specific Spanish regions as is the king, and thus may be less relevant to other Spanish regions. On the other hand, that cathedral is a site of pilgrimage from a wide region, thus symbolizing transnational connections, and Don Quixote is, after all, not a particularly heroic figure. Also the Italian series of great artworks makes certain definite choices: there are for instance no motifs from Palermo or Sicily. On the other hand, they have historically been appropriated as keystones in a pan-European heritage.

4. *Myths* are used by France and more obviously Greece. There are mythical elements in many motifs, but only in few cases is this presence obvious. The French republican figures of Marianne and the sower are modern myths, once deliberately constructed in order to break with traditional ones. Classical antiquity is explicitly invoked on the two highest Greek coins. The owl of wisdom emphasizes the character of the EU project as an intellectual construction. The abduction of Europe by the bull is a national wet dream, as this virile Greek god conquers his female object of erotic desire. The name 'Europe' means 'the West', and this bull myth connects to a historical process of culture imported from the East. There is thus a potential decentring element in such a self-identification of this continent.[9]
5. *Nature* is depicted in Austria, Finland and Germany. Plants and animals like the Austrian gentian, edelweiss and alpine primrose, the Finnish flying swans and cloudberries, and the German oak twig all offer ambivalent implications. On one hand, they may contribute to a naturalization of nationalist constructions by illustrating a kind of Blut-und-Boden philosophy of people, nations and cultures bound to the very soil of a specific geographic area. On the other hand, nature rarely respects political boundaries: swans are nomadic, migratory birds. *Sound of Music* famously made the song 'Edelweiss' a prototypical symbol for the Alpine region, but not only is it hard to distinguish Austria from non-EU Switzerland in this respect: the film was also a typical Hollywood product and the plant can be found lots of

places. A German coin tradition has used the symbolism of oak groves as ancient places of Germanic worship, but oaks are traditional symbols also for Zeus, Jupiter and Kybele, as well as in Christian, Jewish, Indian and Chinese myths.

The boundaries between these main generic types are fleeting. National symbols may integrate natural or mythic elements. National symbols may have transnational or even global aspects, whereas plants or animals also have shifting links to a specific national soil.

Groups of countries

One may tentatively discern four main groups of countries, depending on the general and dominating patterns in their euro coinage.

1. *Nationalists*. Half of the twelve main euro countries lean towards the national side, representing themselves by symbols that primarily point out their specificity in relation to European neighbours. It is the monarchies that have generally taken this road, showing the faces of their kings and queens, but there are a few exceptions – in both directions. The three BeNeLux monarchies all depict their monarchs on all their national coin sides. The Iberian peninsula offers interesting exceptions. The Spanish monarchy has its king only on the largest value coins, and I will therefore place it in another category. On the other hand, Portugal is nowadays a republic, but still has gone the nationalist way and chosen to use the old royal seal and coat-of-arms by the birth of Portugal as a nation in the twelfth century – not a royal head but still a royalist form of national symbol.
2. *Universalists*. France and Ireland have national symbols that invite global interpretations of a much less separatist kind than the previous nationalist ones. Republican symbols are integrated parts of a universalistic discourse and practice, expressly appealing to supposedly universally applicable human values. The figure of the sower is associated with divine creativity and human culture in general, perhaps also to the Christian Sermon on the Mount and thus to missionary activities, but primarily secularized ones in the spirit of Enlightenment, ambiguously linked to colonialism. The harp makes a non-verbal claim of a similar kind, building on the force of instrumental music to move hearts across linguistic and national boundaries. Again, this can be criticized as an ideological illusion, covering the fact that musical life fuels divisive borders between people or cultures, only along different lines than those of verbal culture. Still, the harp does not have a fixed semantic link to any particular territory or state apparatus, at least not to those EU citizens who are not very well informed about Irish mythology, and it may therefore be seen as a kind of universalist statement.
3. *Culturalists*. On Austrian and Italian coins, cultural history clearly dominates. This may be a way to boost one's own grandiosity by claiming copyright for the treasures of cultural creativity in historical heritage. Anyhow, the effect is one of historization and culturalization. Human artefacts from various epochs are lifted up as crucial for collective identification, implying at least a potential for relativization of values. Pointing at aesthetic perfection as the ultimate key to values puts more dangerously divisive political issues in the background, in favour of taste issues that may certainly be controversial but usually less violently so. This is particularly true for the most classical of subjects, but due to the way that art history tends

to de-politicize and universalize artworks, even the Boccioni image of movement, for instance, is easily appropriated as a kind of UN-protected 'world heritage', in spite of the somewhat problematic nationalist war cult of some of the proponents of Italian Futurism. Also the more political persons and buildings chosen by these two countries tend to emphasize peaceful and cooperative efforts rather than national separatism, notably Bertha von Suttner. A curious exception is the Marcus Aurelius statue, since it originally stood on the column in Rome that was inaugurated in the year 193 to commemorate the victory of this emperor over the Germans. However, even this and all his other martial deeds are today easily forgotten in favour of his reputation as a noble and self-reflecting secular thinker, depicted in that famous statue as a prince of peace. Another one may be the Coliseum, where many European slaves to the Roman Empire were once mercilessly slaughtered. Yet none of these motifs are tightly knit to any particular national project, since they mostly predate the late birth of Italy's modern nation state. Many artefacts and buildings have been created through the exploitation of foreign workers or cultures, but the ones chosen in these cases now seem not to exclude transversal identifications. Being included in heterogeneous series, they show artefacts from different historical epochs as a rather arbitrary chain of gems that could be wilfully extended by others, with a slightly different meaning, adding to the historicity and thus secularizing relativity of culture.

4. *Chameleons*. As has been argued here, most motifs have certain potential for ambiguity – being interpretable in different and sometimes oppositional directions. Some nations present themselves in series of images of highly divergent kinds, combining the previous positions and adding yet others. Thus, Finland, Germany, Greece and Spain use similar national symbols as the first groups (royalties, heraldic animals and coat of arms), but mixed with efforts to transcend borders by adding consciously transnational motifs, either culturalist or naturalist ones. Added to this, many of their chosen motifs are often in themselves ambiguous. Take for instance the Greek Velestinlis-Fereos who was an intellectual and a creative poet but also an activist of national liberation, and all three of the Greek individuals depicted combined national liberation from some foreign power (that is, from Ottoman Turkey) with coalitions with other European countries. If Turkey eventually joins the Union, this separatist symbol will seem, to some extent, to run against the main rhetoric of peaceful collaboration between the member states. Likewise, the Greek ships combine many different functions, from classical Mediterranean trade cosmopolitanism over warfare vessels to global oil distribution. And with mythical subjects on top of this, Greece certainly presents a quite open and ambivalent series. Finland and Spain likewise combine national symbols with cultural or natural themes with transnational implications, as has already been discussed. Germany is an equally divided case, with the dark heraldic eagle and the oak twigs framing the intermediary motif of the Brandenburg Gate which is itself extremely ambiguous. It is a symbol of German unity, from Prussia to the reunited Bundesrepublik of today, but it also invokes first the struggles between Germany and France and then the Cold War divide between East and West Germany. The official explanations of this motif repeatedly stress this tension, emphasising that from having been a celebration of anti-French war and then a heavily fortified point of division, it has today become a gate for intense crossings. This is underlined by the chosen perspective, emphasising the road through the gate rather than the wall in which it was previously a closed door. As Gerard Delanty

(1995, p. vii) optimistically states, 'Berlin is no longer the symbol of a divided Europe but the capital of a united Germany'.

All categories often blend, as for instance even the most innocent flower is apparently chosen for its associations with a national identity, and the boundary between mythology and nature is permeable. Many of these multi-faceted national symbols have developed historically in fierce struggles against other (surrounding) nations, although in some few instances there are implications of some kind of inter-European cooperation. In all, there is a slight tendency to a north/south division, with wider sets of images from down south than from the Lutheran, and possibly more iconoclastic, north. This pattern is superimposed on and partly coincides with a political differentiation between constitutional systems – monarchies and republics – most of the remaining monarchies today being found in the north. The age and historical experiences of each national formation also affect its numismatic style. Simple generalizations are hard to make. William Johnston (1991) argues that national differences in forms of celebration can be related to a kind of 'civil religion' used to justify and legitimate the various regimes. In France, the French Revolution is always in focus. Germany has a 'civil religion of *Kultur*' with cultural personalities in focus: artists, philosophers, musicians and writers. Austria relies on the music and theatre of the Hapsburg Empire. Italy favours Catholic saints, cities and regions over the nation as such. Britain's civil religion circles around the monarchy. Johnston (1991, pp. 52ff) sums this up in a main dichotomy between a French and a German model, stressing either political ramifications or apolitical creativity. This is compatible with the euro coins.

Traces of transition

Most countries choose stability rather than innovation in their euro designs, leaning heavily on their conventional range of symbols, but there are some exceptions. The German Brandenburg Gate expresses a historical transition from division to unification. Spain and Portugal both gave up the usual themes from their old colonial history, which might have been problematic in relation to the European project. Classical colonialism was a violent competition between European states, which contradicts the present efforts of peaceful cooperation. The colonial imperialism in the third world certainly resulted in strengthened global interconnections, but in an extremely unequal and coercive manner that is hardly good marketing for Europe in relation to Africa, South America or Asia today. Their old motifs showed men who opened up the world for Europe's exploitation, undoubtedly with many gains but at the cost of so much blood, human suffering and uneven economic exploitation that it must be considered as absolutely one of Europe's most problematic contributions to world history. While Portugal retracted to a more inward-looking nationalist stance, Spain – singular among traditional monarchies – dared to expand its image in transnational and even self-ironical directions, including artistic, architectural, literary and religious themes in its self-image. However, this modernizing tidying-up effort conceals the colonial aspect of Europe's history that has been essential to its very formation and self-understanding as a continent in contrast to its others.

Some nations have made selections and minor refinements that underline common European values and inter-national links, thus showing how each country contributes with its own voice while interplaying with the surrounding others. The Austrian, German and Finnish plants have some regional specificity but may also allude to the issues of global ecology that are one of the reasons for transnational cooperation. Finland lets aggressive heraldic lions be accompanied by migratory birds that know no boundaries and may symbolize the late modern age of mobility. Buildings and artworks have been crucial to the history of each country, but also for international relations and visiting foreigners. Many of the depicted individuals have been cosmopolitan in their lives and work, and are well known across the continent. The French republican themes intend to unify the world, and da Vinci's Vitruvian man has a similarly universal intent in signifying the Renaissance focus on humanity abstracted from all characteristics – except gender, where masculinity continues to rule.[10] And while the German Brandenburg Gate has a painful history of division, the reopened road running through it gives hope for new encounters between East and West. The BeNeLux and other monarchies have given more meagre contributions to this process, reducing their collective identifications to one single and in practice rather marginal aspect.

Unity

The whole set of euro reverse sides displays a striking diversity, but the common obverses and the banknote designs add a coherent direction to the monetary construction of European identity. The cohesion of the union is emphasized by the twelve interconnected stars, all the maps and the banknote bridges, doors and windows that all have an abstract and unspecific character, deliberately avoiding any national or androcentric bias. This abstract unity may be criticized as a magical gesture barely hiding the lack of substantial identity traits anchored in deep-seated popular sentiments of this top-down EU economic project.[11] However, what unites any collective entity is more apparent from the outside than from within. European analysts may overemphasize internal differences and miss common traits. Based on his historical study of European banknote designs, Hymans (2004, p. 24) argues that elements of a European 'commonality may not be out of reach, for the content of collective identities in Europe has been both more *changeable* across time and more *uniform* across space than identity scholars typically assert'. Even the euro is, after all, not a completely empty signifier of European identity.

The dominant image of Europe is as something deeply divided, but striving to overcome internal divisions by mediation and communication, with markets, democracies, civil societies and public spheres as tools. Hence, 'united in diversity' as unity through and by difference. This is one way to read the bridge and door symbols. According to the Dutch writer Cees Nooteboom, 'national identity is itself a melting-pot of cultural influences that transcend nationality and Europeanism consists simply in the recognition of unity in difference' (quoted in Delanty, 1995, p. 129). But not even the banknote designs are innocent abstract symbols for meta-connections. Many Europeans might find them abstract, but, in relation to other continents, there is definitely something typically European in these images. The selection, design and ordering of these anonymous architectural constructions have significant implications. Together, they tell a 'money-story' of two millennia of architectural styles from Roman antiquity to a future-oriented present.[12] This story symbolizes the dynamism and linear progress of western modernity and Enlightenment thinking, where history is conceived as future-oriented progression.

The precise choice of architectural styles offers more signifying cues. The other continents – North and South America, Africa, Asia and Australia – would have made other choices. Only in Europe could the last two millennia be accepted as the appropriate historical totality. The signifying effect of choosing Roman rather than Greek Classical motifs on the lowest-value notes is twofold. Spatially, it avoids placing the origin of Europe to its southeast corner. Temporally, it implies a start around the point zero of modern chronology. Taking a step back behind the magic year 0 would contradict a recurring trope of Europe as a Christian continent, and open the gate to a possibly endless series of previous Neolithic civilizations. Starting with Rome places the origin more centrally in the continent, and coincides in time with the emergence of its dominant religion, which still retains a focal point in papal Rome. With its potentially decentring connections to the Middle East, Athens and Ancient Greece might have implied an ambiguous identity as both European and Oriental, destabilizing the East/West polarity and endangering the self-sufficient idea of Europe as its own product. The symbol, based on a Greek epsilon, retains that liminal origin, but in hidden form, elevating this pre-Christian culture above the mundane flow of history (the money-story) into a universal sphere of pure and eternal foundations. Among national coin reverses, only some Greek motifs go further back than Christian times, reconfirming that all other member states agree to situate the birth of the European project around the year 0 AD in the Roman Empire, whose vast land areas may also seem more appropriate for the claims of a continent than the seafaring group of islands and coastlines that constituted the aquatic network of antique Greece.

Roman culture also fits better with the fusion of engineering technology and humanist ideas that underpins the whole banknote-story. Stabile buildings favour the accumulable (rather than ephemeral) aspects of human culture (rather than nature): fixed rather than variable capital, heritage rather than the fleeting present, products rather than processes, collective rather than individual works, combinations of harmonizing aesthetics and pragmatic technology rather than any other human faculties. Choosing bridges, doors and windows among possible building elements prioritizes infrastructural frameworks rather than meaningful contents, vision among the senses and movement over stasis (e.g. habitation). They are classical pictorial symbols of a typical European dialectics of difference/unity, closure/opening and border/transgression – an urge to mediate, bridge and communicate. The focus on separation in boundary-drawing and border-struggles, the transitions over thresholds in passage rites and liminal phenomena, the current interest in borderlands, hybridity and third spaces – all testify to an obsession with communication across boundaries where Europeans have often been at the forefront – for good and for ill.[13] This aspect of European self-identification can be understood in Habermasian terms as a capacity for communicative action, but also in Foucauldian terms as a power/knowledge effect of panoptical supervision, continued in late capitalist flexibility and surveillance.

It is instructive to consider absences. Potential signs of division are consistently avoided, such as subcultures of all kinds or religious and political symbols, except for the most general and vague ones (like the Celtic harp). There are maps of the EU area, but the decision not to include any flags of member states on the notes or indeed on any of the national coin sides may perhaps be read as a postnational commitment. Nowhere is there any representation of specific countries outside the EU, except for the indirect Greek reference to Turkey as adversary. Norway, Switzerland and the eastern bloc, now gradually integrated into the EU, remain invisible on this first set of coins and notes, as does the surrounding continents and the transatlantic relations that have had such impact on the formation of Europe. The maps on some coins are said to show Europe's place in the world, but this external world remains vague and hidden. There is general talk of openness to other parts of the world, but no specific symbolization of East/West or North/South relations, of European colonialism or American imperialism, besides the microscopic traces of colonial territories left as strangely placed dots on the maps. National symbols are downplayed to some extent (there are no flags for example), but so are specific regions, including those that cross intra-European national borders (like the Basque countries). Women remain marginal, and there is no representation of children or of the working classes. One key feature of modern Europe is particularly absent: mass migration. There are some possible references to border-crossings in the Finnish swans, the Greek independence men, the pilgrimage site of Santiago da Compostela and the pan-European class of royalties, but no clear symbol for the movements of refugees and workers into Europe and between its regions. The euro imagery does not care to represent the new Europe, by excluding any reference both to its recently integrated eastern half and to the many new immigrants from the Middle East, Asia, Africa and South America.

Think of possible alternatives. Natural motifs (plants, animals or landscapes) would be either too specifically bound to one place or too vaguely confined to Europe, and, more importantly, they would not enable a narrative of civilization and progress. Human portraits or situations would again be too specific, but the selection of infrastructural artefacts also has the advantage of hinting at a parallel to the EU as an infrastructural project for communication between nations. Artworks would lack that technological and utilitarian aspect that architecture offers, and which applies so well to the EU, being a tool and a mechanism as well as a work and a symbol. Unlike human beings and some other art forms, the selected buildings are enduring artifices that seem to stand for the stable and trustworthy quality that the Union itself seeks to achieve.

The historical progress told by the paper money-story is thus traced through monumental but utilitarian public buildings, bearing witness to a harmonious combination of aesthetics and technology, and with a practical use for communication purposes. The identifying narrative of the banknotes declares Europe to be a western, Christian unity focused on historical progress, enduring stability, a seamless fusion of aesthetics and technology, boundaries and the processes of communication that cross them.

There are thus aspects of identity, community and unity in this imagery, but more dominant are themes of transport, communication and diversity. So, again, diversity remains the basis for the unity that can be discerned here: as with the EU, the unity of the euro is constructed out of differences. Europe has many historical experiences in common, but belongs to the most internally differentiated world regions, with its old and established nation states, its many divergent languages and its many national and regional myths. Since the end of the Cold War and the fall of the Wall, it does not appear as strongly internally divided as many other continents. It has all kinds of minorities but no longer any clear bifurcation, partly due to the EU project of uniting north and south, east and west. This project joins forces with parallel unifying efforts, such as the ecumenical rapprochement between the Christian churches. Christian religion is a unifying factor, but its role in political and economic life is held back by secularizing counter-forces and by the efforts to better integrate non-Christian minorities, in particular the growing Muslim populations in many states.

Precisely how these new collective identifications of the euro designs will change over time, with the inclusion of more member states and the addition of later editions, is another question. More studies are also needed of how these money signs are read by those who use them, make them and regulate them. The emphasis on abstract forms in a high art formalist style and on images of technocratic infrastructures is typical for the increasingly problematic 'top down' EU project. The Union needs to reconnect to popular images of more specific histories of inter-human and trans- rather than supra-national encounters. Some potential might lie hidden in Euro football and in the Eurovision Song Contest, or more importantly in transnational currents of everyday civic communication and a long history of movements for social justice. But similar traces of interhuman relations and transgressional identifications remain absent in the euro designs. Now, specificity only appears on national coin reverses, where they are still largely carefully confined within nation state borders, with but few signs of emerging transnational and supranational forms of life and identity. Let us hope that future euro editions will open more interesting venues. Still, the euro does offer a unique occasion to study the emergence of a new collectivity – a possibly banal imagined identity but with real effects.

Notes

1. Already as an economic instrument, money mediates between humans, and has been discussed as systemic medium by Marx, Parsons, Habermas and Luhmann (Thyssen, 1991). Its use as a linking device in society implies social functions that have been studied in classical political economy as well as in phenomenological accounts such as Simmel's philosophy of money (Simmel, [1900]1989 and [1896]1991; see also Heinemann, 1969, Müller, 1977, Zelizer, [1994]1997, Buchan, 1997, and Rowe, 1997).
2. European Convention (2003, p. 222). The information on the EU symbols derives from the EU websites (http://europa.eu.int/abc/symbols/emblem/index_en.htm, http://europa.eu.int/abc/symbols/anthem/index_en.htm, http://europa.eu.int/abc/symbols/9-may/index_en.htm); see also the Organization for European Minorities website (http://www.eurominority.org/version/en/devise.asp) and the *Wikipedia* entry on European symbols (http://en.wikipedia.org/wiki/European_symbols).
3. The formulation is from the Schuman declaration (http://europa.eu.int/abc/symbols/9-may/index_en.htm).
4. On the colonial dimensions of 'European citizenship', see Hansen (2000).
5. Billig (1995, p. 41) includes coins and bank notes with flags as normally unnoticed symbols of modern national states that form a kind of everyday 'banal nationalism' that is naturalized and hidden away so that the label of 'nationalism' can be projected only onto 'others'. See also Risse (1998) and Passerini (2003).

6. Facts on the euro designs and launching process derive from the European Monetary Institute: 'Selection and further development of the Euro banknote designs' (http://www.ecb.int/emi/press/press05d.htm); the German Bundesfinanzministerium's website (http://www.bundesfinanzministerium.de/); the ECB website (http://www.euro.ecb.int/en/section.html); Burak Bensin: 'Euro Money!' (http://www.angelfire.com/on/fifa/); an Apple website presentation of how the designs were made: 'Making Money on the Mac' (http://www.apple.com/creative/ama/0201/profile/); Denis Fitzgerald: 'Designing the Euros', *World Press Review*, 2003-01-05 (http://www.worldpress.org/specials/euro/1120web-euro_design.htm). See also Brion & Moreau (2001, pp. 117ff), Ferguson (2001, pp. 332ff), Ludes (2002), Kalberer (2004) and Silveirinha (2004).
7. Hörisch (1996, pp. 13ff) notes the ambivalent double face of money as 'heads and tails', one side with a sovereign portrait (*Kopf*, head) and the other specifying the monetary value (*Zahl*, number), thus initiating a general analysis of the relation between money and poetry, economy and literature, numbers and letters.
8. A considerably extended and illustrated study of all national coin sides, compared with pre-euro designs, is planned for separate publication.
9. Compare similar arguments in Hall (2003) and Rice (2003).
10. Brion & Moreau (2001, p. 51) mention that all famous persons ever selected to appear on banknotes in Belgium, Finland, Spain and Portugal have been male, whereas German and Scandinavian countries have offered women more space.
11. Delanty (1995, p. 128) argues against the 'reifying effect' of the bureaucratic form of EU integration that attempts to 'fashion a European identity using the very tools of nationalism', seeking legitimation in bourgeois high culture.
12. Zei (1995, pp. 337f) describes the narrative told by increasing banknote values as a 'money-story'.
13. For Simmel ([1909]1994, p. 10), 'the human being is the connecting creature who must always separate and cannot connect without separating' – 'the bordering creature who has no border'; to Bachelard ([1958]1994, pp. 222f), 'man is half-open being'. See also van Gennep ([1909]1960), Benjamin ([1982]1999, pp. 494 & 836) and Turner (1969).

References

Anderson, Benedict (1991): *Imagined communities: Reflections on the origin and spread of nationalism*. Revised and extended second edition, London: Verso.

Bachelard, G. ([1958]1994) *The Poetics of Space*. Boston: Beacon Press.

Benjamin, W. ([1982]1999) *The Arcades Project*. Cambridge MA/London UK: The Belknap Press of Harvard University Press.

Billig, M. (1995) *Banal Nationalism*. London: Sage.

Brion, R. & J.-L. Moreau (2001) *A Flutter of Banknotes: From the First European Paper Money to the Euro*. Antwerpen: Mercatorfonds.

Buchan, J. (1997) *Frozen Desire: An Inquiry into the Meaning of Money*. London: Picador.

Delanty, G. (1995) *Inventing Europe: Idea, Identity, Reality*. Basingstoke/London: Macmillan/Palgrave.

European Convention (2003) *Draft Treaty Establishing a Constitution for Europe. Brussels: The European Convention* (CONV 850/03).

Ferguson, N. (2001) *The Cash Nexus: Money and Power in the Modern World, 1700-2000*. London: Allen Lane.

Habermas, J. (2001) 'Warum braucht Europa eine Verfassung?' in *Die Zeit*, 27/2001.
Hall, S. (2003) ''In But Not of Europe': Europe and Its Myths' pp. 35-46 in Passerini, (ed.), 2003.
Hansen, P. (2000) *Europeans Only? Essays on Identity Politics and the European Union*. Umeå: Department of Political Science, Umeå University.
Heinemann, K. (1969) *Grundzüge einer Soziologie des Geldes*. Stuttgart: Enke.
Hymans, J.E.C. (2004) 'The Changing Color of Money: European Currency Iconography and Collective Identity' in *European Journal of International Relations*, 10(1), pp. 5-31.
Hörisch, J. (1996) *Kopf oder Zahl: Die Poesie des Geldes*. Frankfurt am Main: Suhrkamp.
Johnston, W.M. (1991) *Celebrations: The Cult of Anniversaries in Europe and the United States Today*. New Brunswick NJ/London UK: Transaction Publishers.
Kalberer, M. (2004) 'The Euro and European Identity: Symbols, Power and the Politics of European Monetary Union' in *Review of International Studies*, 30.
Ludes, P. (2002) Medien und Symbole: *€UROpäische MedienBILDung. Mit zwei Beiträgen zur Medienzivilisierung von Jürgen Zinnecker*. Siegen: Universitätsverlag Siegen.
Müller, R.W. (1977) *Geld und Geist. Zur Entstehungsgeschichte von Identitätsbewußtsein und Rationalität seit der Antike*. Frankfurt am Main: Campus.
Passerini, L., (ed.), (2003) *Figures d'Europe / Images and Myths of Europe*. Bruxelles: P.I.E.-Peter Lang.
Rice, M. (2003) 'When Archetype Meets Archetype: The Bull and Europa', pp. 77-86 in Passerini, (ed.) 2003.
Risse, Th. (1998) 'To Euro or Not to Euro? The EMU and Identity Politics in the European Union', Arena Working Papers 98/1 (http://www.arena.uio.no/publications/wp98_1.htm).
Rowe, D. (1997) The Real Meaning of Money. London: HarperCollins.
Shanahan, S. (2003) 'Currency and Community: European Identity and the Euro' pp. 159-179 in Passerini, (ed.), 2003.
Silveirinha, M.J. (2004) 'Moeda e construção Europeia: Uma abordagem identitária', paper for the II Congresso Ibérico de Ciências da Comunicação, Universidade da Beira Interior, 22 April 2004.
Simmel, G. ([1896]1991) 'Money in Modern Culture' in *Theory, Culture & Society*, 8(3), pp. 17-31.
Simmel, G. ([1900]1989) *Philosophie des Geldes*. Frankfurt am Main: Suhrkamp.
Simmel, G. ([1909]1994) 'Bridge and Door' in *Theory, Culture & Society*, 11(1), pp. 5-10.
Thyssen, O. (1991) *Penge, magt og kærlighed. Teorien om symbolsk generaliserede medier hos Parsons, Luhmann og Habermas*. Copenhagen: Rosinante.
Turner, V. (1969) *The Ritual Process*. Chicago: Aldine.
van Gennep, A. (1909/1960) *The Rites of Passage*. Chicago: University of Chicago Press
Zei, V. (1995) *Symbolic Spaces of the Nation State: The Case of Slovenia*. Iowa: Communication Studies, University of Iowa.
Zelizer, V.A.R. ([1994]1997) *The Social Meaning of Money*. Princeton NJ: Princeton University Press.

Nation, Boundaries and Otherness in European 'Films of Voyage'

Maria Rovisco

Introduction

A great deal of sociological and historical debate about the idea of Europe focuses on whether a European identity would be able to accommodate varied national and local identities. In public discourse, however, Europe has been forcefully constructed as a matter of 'unity in diversity'...so much so that this formula was recently adopted as the official motto of the European Union (see Sassateli, 2002, p. 440). This discourse emphasizes both the commonalities and particularities of a European culture, from the plurality of national identities to linguistic diversity, from the miscellany of music folk traditions to common cultural heritage in architecture and the arts.

Perhaps not surprisingly, the European Union (EU) has expended great effort in promoting a sense of 'Europeaness' amongst its peoples whilst preserving the idea of cultural diversity across national communities. This has, in large part, resulted from the endeavours of the Council of Europe and the European Commission, evident through a set of symbolic initiatives (for example, the creation and promotion of a European flag and anthem, and the creation of the European City of Culture) directly aiming at creating a sense of common belonging amongst European peoples (Sassateli, 2002). However, the claim of 'diversity' is not inconsistent with the claim of the 'unity' of Europe. The latter commonly invokes Europe's legacy of classical Greco-Roman civilization, Christianity, literary and artistic canonicity, and a moral universalism based on shared values of humanism, rationality, democracy, progress and freedom (see Delanty,

1995; 1996; see also Stråth, 2002). And so the idea of Europe evolved into a naturalized and taken-for-granted notion of a 'geographic space' (for example, the European continent), a 'cultural space' (for example, a space of common culture and identity), and a 'political space' (for example, the European Union as a political entity).

Throughout history, Europe has been defined relationally in opposition to its 'others' whether symbolically embodied by Islamic nations (for example, Turkey), subjugated colonial peoples (for example, African and Asian peoples) or Communist countries. These 'others' are always symbolically placed 'outside' the frontiers of Europe and offer a changing repertoire of meanings against which 'Europe' takes its own significance. As noted by Said (1993, p. 60), 'no identity can ever exist by itself without an array of opposites or negatives: Greeks always required barbarians, and Europeans Africans, Orientals, etc.' Ideological constructions of otherness and the question of Europe's shifting internal and external frontiers have been thoroughly engaged and 'deconstructed' by much scholarship. Historically, the idea of Europe has meant, above all, division, conflict and exclusion, not unity (Morin, 1990; Schlesinger, 1991; Morley and Robins, 1995; Delanty, 1995; Llobera, 2003). In fact, as suggested by Llobera (2003, p. 163), 'for much of its history, the construction of European identity was influenced by the existence of two models, one positive and one negative. On the one hand, Europe meant freedom, democracy, solidarity, rationalism, critical spirit, market economy, etc.; on the other, Europe represented dictatorship, collectivism, passivity, statism, nationalism, etc.'.

Film and, the media in general, are important repositories of publicly available cultural resources to think about individual and collective identities. The media provide, indeed, an important site for cultural analysis because they constitute a major forum for the construction of meaning and, therefore, for struggles over meaning. It is against this background that this paper sets about to probe how the cinematic tradition of European films of voyage reflexively engages, on the one hand, the mingling and interpenetration of distinct local and national cultures which have been typified as European, and on the other, the symbolic boundaries between 'us' and 'them', 'self' and 'other', within and beyond concrete European settings.

The European film of voyage generally fits the thematic, stylistic and narrative concerns of European art cinema which flourished in the aftermath of the Second World War. These so-called New Waves (for example, French *nouvelle vague*, Italian neo-realism, New German Cinema) were tied to state-subsidized cinemas seeking cultural and artistic distinctiveness for their respective national cultures. The voyage genre is also indebted to European literary traditions such as the picaresque genre, the *Bildungsroman* and the *Noveau Roman*, being closely tied to the cultural project of modernity. The latter is based in new modes of interpreting the world and entails a strong notion of the autonomy of human agency (see Eisenstadt, 2001).

Thus the European film of voyage *includes any fiction film in which the journey, thematically and as a narrative structuring device, shapes a tale of self-discovery and social knowledge in contact with otherness* set across European contexts. This cinematic trend spans a period of more than fifty years and includes films such as *Viaggio in Italia/Voyage to Italy* (Rossellini,

1953) and *La Strada* (Fellini, 1954), two iconic films of Italian neo-realism. The acclaimed *L'avventura* (Antonioni, 1959) and *Pierrot le Fou* (Godard, 1965) are also representative of the genre. Wim Wenders' *Alice in den Städten/Alice in the Cities* (1974) and *Im Lauf der Zeit/Kings of the Road* (1976) deploy two fictional journeys, reflecting upon the post-war historical amnesia and legacies of fascism in the Germany of the 1970s. In the post-1989 period, the films by the Greek director Theo Angelopoulos (for example, *To Meteoro vima tou Pelargou/The Suspended Step of the Stork,* 1991, *To Vlemma tou Odyssea/Ulysses' Gaze,* 1995), the 'nomadic cinema' of Tony Gatlif (for example, *Gadjo Dilo/The Crazy Stranger,* 1997, *Exils/Exiles,* 2004), and Michael Winterbottom's *In This World,* forcefully embody the distinctive features of the European 'film of voyage'.

From a cultural sociology approach, this cinematic form offers a useful way to study questions of collective identity because it tackles questions of mobility, displacement, migration and exile within and beyond the European context. The genre concerns new modes of representing the nation by articulating the contrasts between different spatialities – local, national, transnational – as well as the experience of otherness (see Rovisco, 2003). Ultimately, it raises important questions about how the distinction between 'self' and 'other', 'us' and 'them' is constituted. And it does so by critically engaging specific formulations of national identity vis-à-vis questions of boundary crossing and definition.

Theoretically, this chapter shows the need for a more detailed consideration of questions about collective identity formation in relation to questions of boundary crossing and definition. Drawing on the Durkheimian tradition, particularly the literature on symbolic boundaries, it seeks to demonstrate how the idea of the national community is translated into the symbolic boundaries separating 'us' and 'them'. This literature illuminates the principles of classification, symbolic codes, and mental maps which people use to define their own identity and the identity of their community (see, for example, Douglas, 1966; Barth, 1969, 1994; Lamont and Thévenot, 2000; Lamont and Molnár, 2002; Alexander, 2003). Yet scant attention has been given to the study of boundary work, that is, how symbolic boundaries are drawn, maintained and overcome in concrete socio-historical contexts and everyday life settings (see Lamont and Aksartova, 2002).

The approach I use in this chapter aims to bring together some of the analytical tools of the film studies field with the literature on symbolic boundaries. I depart from the assumption that narratives, as cultural tools that we use to make sense of ourselves and social life in general, are made available by the particular cultural, historical, and institutional settings in which we live (Wertsch, 2002, p. 55). Via an analysis of two European films of voyage – *Cinco Dias, Cinco Noites/Five Days, Five Nights* (Fonseca e Costa, 1996) and *To Meteoro Vima tou Pelargou/The Suspended Step of the Stork* (Angelopoulos, 1991), I investigate how through specific arrangements of plot, character and settings these two films deploy a mental map of the national space that is not consistent with the idea of the homogenous nation. At a more theoretical level, I go on to propose the view that the European films of voyage narrative offers a critique of universalistic conceptions of the 'other' whilst avoiding the fallacies of cultural relativism.[1] The underlying question is whether the difference between 'self' and 'other' can be negotiated, even if the symbolic boundaries between 'us' and 'them' remain relatively uncontested.

The films being analysed here relate to two distinct spatialities: (1) Greece as part of the ill-defined and transient 'Balkan space'. This is a spatiality whose cartography has changed widely in an area where different ethnic groups and religions have mingled for many centuries; (2) the Portuguese national space where political borders have remained substantially unchanged since the late Middle Ages, and where the attainment of early statehood and the absence of conflicts with neighbouring countries favoured the model of the homogenous nation.

Boundaries, space and identity formation

The epistemological significance of the notion of boundary is intrinsically linked to our human need to find and impose order on the world, to see patterns, or to make clear distinctions (Strassoldo 1982, p. 246). The idea of boundary exists, in fact, in all societies from primitive to modern (see Llobera, 1994, p. 102). Boundaries have political, cultural and geographical dimensions that influence processes of collective identity formation.[2]

Political boundaries have been comprehended as a limit to the space demarcating distinct political communities (see Goff, 2000). They bear some kind of relation to a territory and to the physical uses of space, but their main characteristic is that they deserve to be well guarded (Nordman, 1998, p. 28). Political boundaries are political constructs, that is, they are artefacts and artificial (Hartshorne, 1938; Baud and Van Schendel, 1997). Historically, they have been associated with fortified zones and are commonly the result of some sort of international agreement.[3] Both in a literal and metaphorical sense, political boundaries are associated with notions of vigilance and surveillance. When we speak of political boundaries as being more or less permeable, we imply that there is some sort of control over the circulation of people, goods and information. This stance also implies that the degree to which this control or vigilance is exerted can change. Political boundaries and border zones perform, in fact, a dual function; on the one hand, they act as barriers or buffers, and on the other, as gateways or bridges (O'Dowd, 2001). Moreover, political boundaries, which are commonly identified with nation state boundaries, are linked to a concept of sovereignty and exclusive control over contiguous territory (Anderson, 1996). Yet, they are not indelible but disputable; often in the aftermath of wars, political boundaries have been redrawn or simply erased. Thus, it is hardly surprising that they easily cut across cultural boundaries and are often at the root of ethno-cultural cleavages and conflicts. In addition, migrants and local communities settled in coterminous border areas are more prone to be affected by the divisive attributes of a political border.

A *geographic boundary* can be identified with a topographic configuration such as a river or a valley, but it needs to be signalled in the landscape with some sort of fortified area (for example, border station) so it can easily be identified. Arguably, some geographic formations (for example, a river) can be transformed into effective political boundaries because they often constitute a natural barrier or obstacle to the mobility of people. It is difficult to erase or move geographic boundaries because they are usually inscribed in the landscape: it is feasible, for instance, to change the course of a river, but not to move a mountain range. The idea of natural border is, however, insightfully contested by Van Gennep who argues that there is no causal link or normal coincidence between natural and political borders because 'the concentration, dispersion and expansion of peoples follows its own logic, and it is not the geographic one, except in rare and passing circumstances' (Van Gennep 1922 cited in Llobera 1994, p. 102). In fact, not mountains, nor rivers, nor seas, nor forests pose an insurmountable obstacle for the movement and expansion of people.

Cultural boundaries can be comprehended as identity markers (or *diacritica*) that enable the distinction between contrasting communities (Barth, 1969). In everyday life people rely on a highly salient and relatively stable set of symbolic boundaries, such as language, religion or ethnicity, in order to make sense of their collective identity as distinct from the identity of other collectives. The content of such symbolic boundaries, or identity markers, is reinvented and negotiated within stories, images and symbols of 'us' that the media constantly launch into circulation, and derive from various sources such as the state apparatus, cultural producers and institutions working within and beyond the state level. In contrast with political and geographical borders, cultural boundaries are not necessarily attached to a specific territory. They can, however, bear a relation with a conception of homeland or other cultural space of belonging, which can be linked either to an ethnic or civic understanding of the nation. Furthermore, cultural boundaries can remain relatively stable and indifferent to the crosscutting pressures of changing social and historical conditions (for example, a dispute over a political boundary). Collective identities are thus meaningfully expressed by symbolic boundaries which people use to separate 'us' and 'them'.

The crossing of a boundary either natural or artificial is symbolically associated with notions of challenge, danger and transgression.[4] As a literary ingredient, the journey has long been associated with the passage from the domain of the familiar to the realm of the foreign. As early as the fifteenth century, the project of European expansion and maritime discoveries transformed the voyage into a movement of detachment from the realm of the 'known' towards the knowledge and comprehension of the newly discovered lands and of the exotic 'other'. In the sixteenth and seventeenth centuries, the epistemological conditions were created to value the voyager as an observer of the exterior world (Matos, 1999, p. 236). In addition, geographic boundaries came to acquire a wider and more figurative meaning in literature, especially in geographic texts, narrative accounts of journeys (for example, travel literature) and travel guides. These literary works played a major role in the description and mental mapping of foreign territories. In her work on the imaginative topography of the Grand Tour, Chard (1999, p. 11) suggests that 'travel entails crossing symbolic as well as geographic boundaries and these transgressions of limits invite various forms of danger or destabilization'. The journey holds here a promise of self-discovery and adventure in the contact with otherness. This sensibility can also be found in the European film of voyage.

The resulting 'encounter with otherness' is not merely a question of going beyond or stepping outside a sharp boundary line drawn along clearly marked geographic or political divides. The encounter with the foreign is rather a matter of crossing boundaries that enables the perception of symbolic distinctions between 'us' and 'them'. These boundaries are usually associated with geographic points or landmarks inscribed in the landscape; Geographic and political markers that would otherwise constitute mere features of landscape or, at most, a suitable barrier to delimit territory are instead laden with a particular meaning by the journey as a trope of discovery. The crossing of a boundary covers here the passage between the space of the familiar and the space of the foreign, that is, the space of 'what is not yet known'. This view implies that the other cannot simply be perceived as the one 'on the other side', for example, the barbarian, the enemy, or one that cannot be known because difference is considered a matter of cultural standards that we cannot comprehend. On the contrary, European films of voyage suggest that the foreign can be made familiar through the act of travelling. As argued by Porter (1991, p. 188), 'in our excursions into previously unknown lands, we discover much which is strangely familiar; and such troubling encounters may destabilize inherited categories as well as confirm them'. The absoluteness of geographic, political and cultural boundaries is unsettled when the journey disturbs the certainties of individual and collective identities, which have long been rooted in the fixities of place and community (see Leed, 1991).

Balkan Voyages: The Suspended Step of the Stork and the limits of community

In *The Suspended Step of the Stork*, we follow the spatial trajectory of a young TV journalist, Alexandre, from Athens to a border-town crammed with refugees. He is working on a story about refugees in Greece when he comes across a man that he believes to be a famous politician who disappeared under mysterious circumstances. As the film starts, we get a hint of the quest that is about to unfold. Whilst the camera zooms in on a refugee's dead body floating in the sea, the journalist recalls in voice-over the 'episode of Piraeus' in which a group of refugees died in the sea after being refused asylum by the Greek government. What is at issue in the film is, then, the interplay of an individual destiny (that is, the journalist) and a collective destiny (that is, the plight of the refugees).

We 'discover' aspects of the plight of the refugees in the border-town 'through the eyes' and unspoken subjectivities of Alexandre. He emerges as a silent and attentive observer of the space he traverses and the people he encounters. Thus Alexandre's personal and spatial trajectory has always to be seen in the interplay with the film's own spatial strategies. Such an unorthodox approach to character (see Bordwell, 1997, pp. 264–265), in pushing the main character to a distant background (Alexandre is mostly framed in full-shot or long shot in a deep focus), not only reduces dramatic tension by disturbing our understanding of the character, but it also subordinates the actor to landscape or décor.

The film deploys two distinct personal journeys: on the one hand, there is the journey of the missing politician to the border-town where he is assumed to be living under the identity of an Albanian refugee; on the other, there is Alexandre's ongoing journey in the footsteps of the missing man. Trying to resolve a man's identity becomes, therefore, a way of getting a deeper insight into the critical situation of the refugees in the border-town. Ultimately, the journey turns into a reflection upon the political and existential overtones of the condition of the refugees.

Alexandre, who begins as a journalist guided by journalistic objectivity, is emotionally drawn into the dramas of the illegal immigrants (see also Horton, 1997, p. 172). However, Alexandre's voice does not reveal the details of life in the town. This role is mostly ascribed to a choric figure, the colonel who patrols the political border. As a privileged observer of events, he explains to Alexandre how so many illegal immigrants and asylum seekers ended up crammed in this northern town on the border with Albania.

From the top of a watchtower, the colonel explains that on the other side of the river that bisects the town there is a village people call the 'Waiting Room'. Through his voice-over we then learn about the growing numbers of refugees from several countries who have crossed the border illegally and are confined to the town whilst waiting for a legal solution to be able to go elsewhere. He adds that this 'elsewhere' acquires for them a mythical meaning. Later on, the colonel attempts to elucidate for Alexandre how the experience of life in the border-town is dramatically different from life elsewhere in Greece: 'Here, in the end of the country everything gains another dimension. Solitude. Uncertainty...a feeling of continuous threat that drives men to madness'.[5] This sensibility is, for instance, powerfully emphasized in the scene where a refugee is found hanged by the neck from a huge metallic loading crane. Whilst the TV crew shoots the scene, the colonel, visibly irritated, tells Alexandre that conflicts amongst refugees erupt everywhere. He confesses hopelessly that he cannot understand what is going on since no one says a word: 'They crossed the border to find freedom and now they create new borders here! They split this slum making the world even smaller! To make things worse not even a word! It's the law of silence!'

The Suspended Step of the Stork is particularly concerned with the impermeability of the Greek/Albanian political border in the Epirus region, symbolically reflecting much of the tension underlying the West and East divide. Furthermore, throughout the twentieth century both countries have endured rivalries over disputed parts of this region. Against this backdrop, the film engages the divisive effects of the political boundary vis-à-vis the physical presence of the border station.

A landscape pervaded by watchtowers and military personnel plays a very significant role in depicting a space where people's sense of fear and displacement is intrinsically related to the visual proximity of the political boundary. The river that bisects the border-town offers an important geographic marker for the political border. It helps to emphasize the sense of separateness from Albania where the refugees who did not manage to cross the border are still waiting on the other side. As I have suggested, those who live close to the political border are more prone to be affected by its divisive effects. In this case, the political marker could not be more suitably connoted with notions of danger and transgression. The physical border limits a territory of despair (Pierron, 1995, p. 143). This is strikingly illustrated by the film's visual imagery that, whilst commenting on issues of social exclusion affecting the refugees, shapes the image of the border-town as a true 'no-man's land'.

The constant portrayal of jeeps and military personnel, the visual prominence of watchtowers, the image of a man hanged on a huge metallic crane surrounded by a group of howling women, the fight initiated at the café, the bleak and often completely deserted streets of the town (either wet or covered in snow), the slummy and ruined exteriors, the depiction of boxcars serving as temporary houses, all help to emphasize the climate of uncertainty, despair and strict surveillance surrounding daily life in the town. Moreover, homey and familiar environments (for example, cosy houses, schools) are rarely, and never clearly, depicted, thus stressing the strangeness of the place. The space of the border-town is often perceived as imprecise and unconfined due to an unorthodox use of lighting that darkens settings and conceals characters. Most scenes take place in public places (for example, the marketplace, the hotel hall, cafés). Even the orthodox wedding celebrated by the border-river is displaced from its traditional religious setting.

The space of the border-town is depicted, above all, as a space where everyone feels estranged. Even a border patrol officer tells of his sense of displacement and loss in consequence of life in such a place: 'I'm a tragic character. I'm paid to watch over the border. My wife is in Athens and my daughter in London for her alleged studies. The gypsies! Who knows where I'm going to be sent tomorrow!' We get no sense of community life in the town. Relations between people are marked by fear and suspicion, and tensions and quarrels between the refugees go unresolved due to a striking silence that prevents police intervention. Thus we need to ask whether a group of refugees can ever be described as a community (see Kelly, 2003). In this particular context, people appear to be bound to an in-between situation in a visibly alienating space. It seems therefore almost impossible to think of the group of refugees as bound by close kinship or solidarity ties. Moreover, the overt and latent conflicts amidst the refugees suggest that there are strong differences both amongst the refugees themselves and between 'them' and the locals. When the reporter asks for information about the whereabouts of the man he believes to be the missing politician, he is told to look at the refugees' quarter or in the café where 'they' usually go. Alexandre also learns that people in the village relate minimally to 'them'. There is a clear indication that the rivalries between refugees are accompanied by discrimination from the locals.

In *The Suspended Step of the Stork* divisions and barriers to intercultural understanding between distinct individuals and groups (for example, family, ethnic group, a group of refugees) are mostly a consequence of boundaries that are both internal (that is, internalized and psychologically experienced by an individual) and external (that is, imposed by a highly impermeable political border). Feelings of fear and repression reigning amidst the refugees are reinforced by the presence of the border-station and the dangers associated with the crossing of the political border. The inhumanity of this border, which clearly emerges as a barrier to the mobility of people, is powerfully enhanced in the way many refugees risk their lives to smuggle a pack of cigarettes or a music cassette. When Alexandre visits the border-town for the first time he is taken by the colonel to the riverbank where they both catch a man retrieving, in a tiny raft, a cheap cassette tape recorder that is playing music. The colonel takes the opportunity to warn the man of the risks of smuggling. The 'real' porosity of the border is also illustrated in the way the refugees who manage to cross attempt to maintain contact with those who stayed 'on the other side', that is, in the village known as the 'Waiting Room'. This issue is also addressed in the scene set at a traditional Greek café where Alexandre is arranging with the TV crew to shoot the wedding by the river on the following day. The colonel tells Alexandre that 'after they [the refugees] cross the border they meet secretly once every year regardless of the danger. Often we find bodies floating on the river.' This is a striking example of how kinship bonds and loyalties defy the 'deadly' political border. This is also to say that cultural boundaries can run strikingly at odds to the divisive attributes of an 'artificial' political border.

On the one hand, then, the sense of confinement associated with the border-station produces increasing divisions amongst the refugees in the border-town; on the other, it is also clear that enduring loyalty and kinship ties prove resistant to the separateness imposed by the political border. In the last instance, the absurdity of this highly impermeable political border between Albania and Greece becomes an even more vivid reality since throughout the film we are made aware that one single step over the border means death. This is forcefully shown in the scene in which the colonel lifts his leg like a stork by the blue line that marks the political border with Albania (a gesture Alexandre repeats in the closing scene) whilst saying to Alexandre: 'If I give one more step, I'm on the other side and I'll die'.

In such an alienating place, identity clearly emerges as constituted against the figure of the 'threatening Other' (Castles, 2003, p. 23). Implicitly, every individual may be potentially constituted as an 'other' to his or her fellow neighbours. Understandably, people's daily lives are affected by fear, subjugation and the insecurity of having nowhere to go. It is clear that the group of immigrants involuntarily retained in the border-town by the Greek government hardly fits the ideal of community as a closely bonded collective of people united by shared values and culture (*Gemeinschaft*). Thus identity here is not only potentially shaped in conflict with 'the other' but also in the absence of a clearly demarcated 'us'. In the impossibility of achieving a sense of collective belonging, the boundaries between 'us' and 'them' may well become blurred for each of the refugees inhabiting the town. Simmel's (1970) figure of the stranger works well here to illuminate the all-embracing condition of the refugees. The stranger merges proximity and distance in an ambivalent constellation, finding herself caught in-between two groups to which she does not entirely belong (see Vandenberghe, 2005, p. 125). *The Suspended Step of the Stork* succeeds, after all, in suggesting that in an alienating and 'in-between' border-town no one can, in fact, feel at home. And such a sense of displacement finds expression in an all-encompassing nostalgia for a 'place where one can feel again at home': 'How many borders do we need to cross to get home?' says the allegedly missing politician in his 'new' identity.

The film implicitly critiques Greek immigration policy especially with regards to the issue of illegal immigration that was particularly fiercely debated at the time the film was released in 1991. Both narratively and visually, the plight of the refugees is related to the Greek state's incapacity to deal with issues of illegal emigration. We are confronted with the brutal fate of the refugees who are refused legal papers in order to start a new life. Furthermore, the fact that the missing politician is allegedly assuming the identity of an Albanian immigrant draws attention to issues of discrimination affecting Albanian immigrants in Greece. As suggested by Lazaridis and Wickens (1995 cited in Lazaridis, 1996, p. 345), the prejudice and xenophobic treatment that is applied to many illegal Albanian refugees (negatively cast as primitive and untrustworthy) is in part related to the fact that the Greek government wants to use the issues of illegal Albanian immigrants in negotiations with Albania about the human rights of ethnic Greeks in Albania. Subsequently, xenophobia towards Albanian immigrants becomes an even more complex issue when we consider that discrimination can hardly be made on grounds of ethnicity or religion. Albanians have, in fact, a genuine claim of Greek ethnicity (Lazaridis, 1996, p. 345).

It is useful to briefly recall here that, with the establishment of a Greek state in the early 1830s, the majority of the Greek population was left in Ottoman lands beyond the new nation state's political borders. This gave rise to an irredentist ideology known as the 'Great Idea' that encouraged the so-called 'map mania', which aimed to legitimate Greece's desire for territorial expansion. 'Map mania' reflected a wider recognition of the political role of geography in laying claim to the territories encompassing all the eastern lands inhabited by the Greeks in a reconstituted Byzantine empire[6] (Peckham, 2000; see also Roudometof, 2001). At the same time currents of nationalist thought were defining Greek identity primarily in ethno-religious terms. As argued by Peckam (2000), the naturalization of political boundaries in the late nineteenth century and early twentieth century was in part a consequence of increasing links being drawn between biology and territorial identity. This cultural and political historical legacy is subtly embedded, though never directly addressed, in the narrative of *The Suspended Step of the Stork*.

Arguably, then, the Greek TV reporter's quest for the missing politician, living under the identity of an 'other' (that is, an Albanian asylum seeker) encompasses a search for what it means 'to be Greek' in view of current social and political changes at the close of the twentieth century. This sensibility is forcefully hinted at when, shortly before embarking on his journey-quest, Alexandre quotes to his girlfriend the concluding lines of the missing politician's influential book: 'And what are the key words we could use in order to make a new collective dream come true?' (as quoted in Horton, 1997, p. 166). This is a time when Europe becomes the stage for a rather unpredictable making and breaking of nations in the Balkan region, the expansion of democracy to the eastern Europe and Russia, and new patterns of mass migration into western Europe originating mostly from eastern Europe and the Middle East.

In short, *The Suspended Step of the Stork*, a film set in the socio-historical context of Greece of the early 1990s, critically engages the exclusionary practices and rhetoric of the modern nation state in a region that has historically marked the encounter of different religions, ethnic groups and civilizations. Alexandre's journey-quest for a missing politician produces an awareness of the condition of the refugees in Greece and, as such, it plays up the significance of the symbolic boundaries of the Greek national community. We have seen how the symbolic boundaries separating 'us' and 'them', translated into the highly impermeable political boundaries, are challenged in everyday life settings by the refugees. Regardless of the risks and prohibitions associated to the 'guarded border', the refugees defy the latter for the sake of maintaining strong kinship and loyalty ties and, more importantly, a sense of community.

Experiencing the nation in *Five Days, Five Nights*: between the local and the national

Unlike *The Suspended Step of the Stork* that takes on the condition of the refugees in Greece of the early 1990s, *Five Days, Five Nights*, released in 1996, looks at the recent past of the Portuguese dictatorship. This film fits the purposes of this chapter in three ways: firstly, it is a good instance of 'deconstructing' the idea of Portugal as a culturally homogenous nation; secondly, it is also a good case for questioning cultural understandings of contemporary Portugal as 'European' and 'democratic' vis-à-vis its recent antidemocratic past; finally, the historical conditions prompting the formation of the Portuguese national space (for example, the attainment of early statehood, absence of significant ethno-cultural minorities since the late Middle Ages, stability of political borders) offer an interesting counterpoint with those which characterized the emergence of the Greek nation state (for example, late statehood, constant boundary-drawing, ethno-cultural diversity).

In *Five Days, Five Nights*, we follow the aimless trajectory of André, a political escapee, across late 1940s northern Portugal. The film is adapted from a novella[7] by the historical leader of the Portuguese Communist Party, Álvaro Cunhal. *Five Days, Five Nights* concerns primarily the interplay of André's journey of escape with the 'discovery' of a remarkable cultural enclave in a remote borderland in northern Portugal. The story is set at the height of the Estado Novo regime whose ultraconservative, anti-Liberal and Catholic ideology resisted modernization and secularism. André is a young and idealistic political prisoner who escapes from prison during the isolationist times of the dictatorship in late 1940s Portugal. For five days and five nights he wanders, across mountains and valleys, around the barely accessible northern region of Trás-os-Montes (*Behind-the-Mountains*) that borders Spain. Sometimes alone, sometimes not, he is helped illegally to cross the border by the enigmatic and seemingly untrustworthy Lambaça, a smuggler of notorious reputation. What keeps André on the move amidst the ceaseless danger of being caught by the political police, and his mistrust in Lambaça, is his overwhelming desire to abandon the country and the certainty that there is no way back home.

As the voyage goes on we become aware that André is not, however, rejecting the idea of the nation as a homeland, that is, the idea of Portugal as place that contributes to the constitution of collective identity through a territorially based community (see Entrikin, 1999). What we find, paradoxically, is a 'journey into society' since, in the process of abandoning the country, André gets a deeper insight into the textures and limits of the Portuguese national identity. Meanwhile, the journey also evolves as a typical 'journey of initiation'; through his contact with Lambaça and several 'underworld' characters, André gains self-awareness and learns about 'others' in a manifestly foreign space. This aspect is also suggested by the fact that the story is told from André's point-of-view. It unfolds in a first-person narration as a recollection of events situated elsewhere in the past and filtered through André's personal memories. The film starts with the following lines in André's voice-over: 'When not yet nineteen-years old I saw myself forced to emigrate. I was given money, someone's address in Oporto, and I was told that there everything would be arranged regarding the crossing of the border to Spain'.[8]

A sense of home is here tied to André's need to make familiar, or to comprehend, the values, identity and culture of a nation that is being 'discovered' through the lens of its intrinsic cultural diversity. Like in a *Bildungsroman*, *Five Days, Five Nights* goes on to reveal a world that is meant to acquire the comforting dimensions of familiarity (see Moretti, 2000, pp. 34–35). This is why the film is not only telling a story about a young man's escape from an oppressive political regime, but also a story of social learning. The story evolves, then, with a suggestion that André's increasing awareness of the need for a change of the existing political system and dominant societal values is linked to the experience of the journey itself. André's optimism and political militancy is counterpointed in the stance of Lambaça, his journey companion, a cunning and experienced smuggler with no apparent political convictions. This is powerfully conveyed by Lambaça's emblematic words: 'I know how it is. When one is young one thinks one can change the world. But nothing changes, whatever happens.' Such a position could well stand as an expression of the political apathy of a disfranchised rural population (see Martins, 1971, pp. 83–84).

The fact that, in *Five Days, Five* Nights, the journey is set against the backdrop of a geographically isolated and impoverished region of the country allows the film to document the way of life of local rural communities at the time of the dictatorship. In following André's personal and spatial trajectory, we learn about the life of people bound by strong ties of solidarity and forced to survive at the cost of smuggling in the barely accessible province of Trás-os-Montes. Through André's voice-over we learn his inner feelings and thoughts about the experience of the journey. A mix of uneasiness, distance and curiosity marks his interaction with the group of smugglers who are helping him to escape to Spain. His words conveyed in voice-over leave no doubt that he feels like a stranger: 'I felt profoundly disturbed in recalling how I was espied and gazed on by those glances. I didn't know that universe charged with silence and mumbled speeches.'

Like Alexandre in *The Suspended Step of the Stork*, André visibly embodies the voyager travelling a seemingly familiar terrain – in this case the Portuguese national space – which is being experienced as strange. Both voyagers emerge as silent observers of a 'reality' they struggle to comprehend. But unlike Alexandre, André is not being confronted with a 'no-man's land' as is the case of the Greek border-town where a sense of community amongst the refugees is prevented by fear and uncertainty. In contrast, *Five Days, Five Nights* unfolds the fine details of a specific 'border-culture' characterized by a strong sense of community, very much in the sense of Tönnie's concept of *Gemeinschaft*. It is 'through the eyes' and subjectivities of André that we have access to aspects of life of a 'border-culture' that finds little expression in the idea of the homogenous nation conveyed in the Estado Novo's official narratives.

Five Days, Five Nights succeeds, after all, in drawing attention to a strong 'border-culture', fostered by the illegal activity of smuggling, which in no way resembles the group of refugees crammed in the Greek border-town. Smuggling, as a predominantly nocturnal and pedestrian activity, represents an important local source of economic subsistence for those inhabiting the region, especially in the difficult times of the dictatorship. Settings and landscape are depicted in a naturalistic fashion that suggests the sense of isolation of people inhabiting this predominantly rural area. Shortly after the beginning of the journey, Lambaça and André get off the train at a small railway station at a lively market. Whilst they mingle with locals at the busy market so as to hide from the political police, we are shown the details of a world in which people in traditional peasant costumes sell vegetables, chickens and ducks. Such 'realistic' visual imagery reveals the local economic and cultural practices of a peasant way of life. Visual details such as the absence of electric light in interiors, the shabby and scarce furnishings, the overemphasized presence of farming tools and products and a naturalistic use of lighting that strikingly darkens settings, suggest the state of destitution and the hardships of life of an isolated rural population.

We also become aware of the strong ties of solidarity and trust governing relations within this 'borderculture'. This is particularly apparent in the interaction between Lambaça and his smuggler friends. Dialogue is scarce, and relationships appear to rely on a mutual understanding that goes beyond words. Such strong kinship ties are even more compelling if we consider that the film is set at a time when strict police surveillance was imposed by a repressive state apparatus, as in the scene where the political police subject the family of smugglers to a brutal interrogation. Despite the threats, they do not provide information of the whereabouts of André and Lambaça. The generosity and kindness of the prostitute Zulmira, who refuses to set a payment agreement after offering shelter and food to both André and Lambaça, is another example of the altruism governing the relations among people. These social practices can be understood as an expression of what Santos (1994) considers a strong 'welfare society' which in the Portuguese case replaced a deficient welfare state. For Santos (1994, p. 64), the 'welfare society' comprises those networks of face-to-face relations and mutual help based in kinship and communitarian ties. The 'welfare society' is organized according to traditional models of social solidarity, where small social groups exchange goods and services on a non-commercial basis under logic of reciprocity.

As in *The Suspended Step of the Stork*, we are faced in *Five Days, Five Nights*, both narratively and visually, with the question of the impermeability of political borders. At first glance, the film appears to be concerned with the almost complete lack of permeability of the Portuguese national borders in a time of severe restriction of the circulation of people and goods across the frontier. This is mostly suggested through a depiction of the way of life of Lambaça and his smuggler friends in the inaccessible enclave of the Trás-os-Montes region. At a more figurative level, however, the film brings up important questions about the artificiality of political borders. This is in part activated by powerful imagery and the use of specific settings (the film is shot on location in a neo-realist fashion). In contrast with *The Suspended Step of the Stork*, deep focus cinematography, 'continuity' cutting and the moving camera (commonly found in classical Hollywood cinema) foreground a rich visual imagery. Visual depictions of Lambaça and André, either walking or overlooking high slopes and small canyons prove effective in depicting a geographic area that for centuries has offered a boundary zone for Portugal in the northeast.

If the story line follows the flight of a young political prisoner across the Trás-os-Montes region until he successfully crosses the border, the visual imagery helps to deploy a national border that is more artificial and composed than its geographic outlines appear to suggest. Certainly, the rocky mountain slopes and dangerous canyons Lambaça and André have to cross offer a 'natural' barrier to the mobility of people and make the Trás-os-Montes area a suitable border zone. There is not, however, any visual indication of a border station or any other sort of fortification to signal the border in order to prevent the unauthorized movement of people and goods. This is made particularly visible in the last scene. When Lambaça decides to inform André that they have already crossed, André reacts with surprise and disbelief. Although he must have suspected already that the voyage was coming to an end, there was no border station or even a signal post that could function as a significant marker of the political border. In the end, the crossing of a small and quiet stream of water when they were already on Spanish soil symbolically marks the moment of the 'passage' to Spain.

This closing scene hints that national borders are, above all, a product of ideological manoeuvres that are likely to make use of relevant geographic formations, such as rivers and mountain ranges, to consolidate and legitimate the nation's political outlines. In fact, distinctive terrain, islands, mountainous areas and peripheries have always helped to emphasize differences and secure the recognition of a national territory (Williams and Smith, 1983). What is also apparent, however, is how the 'invisibility' of the political boundary contradicts one of the cornerstones of the so-called modern geopolitical imagination, that is, the assumption that the world is divided into distinct sovereign states circumscribed by precisely defined geographical boundaries (O'Dowd, 2001, p. 96). The film also makes settings and narrative work together to subtly unveil the topographic similarity between Portugal and Spain along the border area.

We have also to consider the relation of a specific border zone (with a distinct border-culture) vis-à-vis the state apparatus. In this sense, *Five Days, Five Nights* suggests that geographic and economic isolation in the Trás-os-Montes region may well have fostered a locally bounded 'border-culture', which benefited from the proximity of the political national boundary to assure survival in the harsh times of the Estado Novo regime. This also means that, in the Portuguese context, the locale offers a better ground for collective affiliation than the idea of the territorial nation. This argument is stronger if we accept that in the Portuguese case we are, in fact, dealing with a highly heterogeneous culture that is difficult to assimilate within the notion of the homogeneous nation. As suggested by Santos (1994, p. 133), in Portugal, the state never succeeded in promoting a strong homogenous culture capable of differentiating a Portuguese national culture from other national cultures. Thus both the 'local' and the 'transnational' (through the idea of a Portuguese imperial nation composed of far-flung parts worldwide) have been much stronger than the national in defining a Portuguese national culture. Only very recently has the democratic Portuguese state used cultural policy and propaganda to promote a true homogenous national culture (Santos, 1994, p. 136). This has to be seen as a strategy for strengthening Portugal's position in the European arena.

Furthermore, *Five Days, Five Nights* illustrates the opposition between a developed, and highly populated 'littoral', and an underdeveloped, rural and deserted 'interior'. This symbolic distinction became a distinctive mark of 'modern Portugal' from the late 1960s onwards. As argued by Ferrão (2002, p. 156), the boundary separating the 'interior' and the 'littoral' translates as the subjective delimitation between the 'included' and the 'excluded' in modern Portugal. This boundary is articulated not as a precise cartographic distinction, such as that coding the outlines of the national territory, but through the idea of a 'mental map'.[9]

As powerfully hinted at by the film, the local and the rural, here embodied in life in the Trás-os-Montes region, better translate a generalized sense of exclusion. This subjective mapping of the country codes the experience of living in the 'interior' with a lack of employment and education opportunities as well as with poor regional development. These societal and economic developments (for example, poor industrialization, lack of an entrepreneurial bourgeoisie, political resistance to the modernization of agriculture) have been in part the result of structural socio-historical conditions that go back to the authoritarian regime and continue in the democratic present (see Rosas, 1998). In the 1990s, this picture has changed slightly with the opportunities of regional socio-economic development created by the European Union, the improvement in the networks of transportation and communication and the expansion of common patterns of consumption (Villaverde Cabral, 1992). But these developments have so far had little impact on local and national territorial identities (see Ferrão, 2002). *Five Days, Five Nights* forcefully plays up the significance of the social and economic asymmetry of the national culture, whose complexity is still, today, often dismissed in both the political and the media agendas. Interestingly, many small rural populations today, especially in the Trás-os-Montes region, closely resemble those one could find in the authoritarian Portugal of the late 1940s. Nowadays, and despite steady modernization, the unwavering 'reality' of geographically isolated populations (see Peixoto, 2002; see also Fortuna, 1997) constitute a threat for the advertised image of 'modern', 'cosmopolitan' and 'European' Portugal. Hence, there is ground to argue that the local and the transnational have proved through time (history) and space (culture) stronger than the national in defining Portuguese national identity. *Five Days, Five Nights* shows that the way people experience the Portuguese nation through their attachment to local culture constrains their imagination of cultural constructions such as a unified national culture and an imperialistic nation, which are both enforced by the Estado Novo ideology.

Conclusion

In deploying the journey as an act of crossing boundaries, the two European films of voyage in this chapter reflect upon the symbolic ways in which political, geographic and cultural boundaries shape and inform the process of collective identity formation within and beyond the nation in concrete socio-historical contexts. In doing so, the films articulate the connection between the past and present of historically rooted communities, bringing to the fore issues of both suppressed history and official history, forgotten memory and collective memory.

Via specific arrangements of settings, plot and character, *The Suspended Step of the Stork* and *Five Days, Five Nights* reveal some of the ways in which the act of crossing a boundary meaningfully associated to geographic and political markers plays up the significance of the symbolic boundaries between 'self' and 'other', 'us' and 'them'. In this way, I have shown how European films of voyage articulate some of the ways in which people challenge these very same boundaries in concrete historical moments (time) and everyday life contexts (space). This is also to say that both films reflect upon the experiential dimension of national identity; that is, the ways in which individuals subjectively experience their nation through the act of journeying and crossing boundaries. This aspect of collective identity formation is often neglected by constructivist approaches that focus on processes of cultural fabrication of the nation by local and national elites. Ultimately, by taking on the journey as a narrative device and a trope of discovery, this chapter shows how the European film of voyage narrative proves a useful heuristic device to explore the connection as well as the tensions and clashes between 'self' and 'other', 'us' and 'them'. It does so by challenging the dimension of familiarity that is commonly ascribed to categories of national imagination.

Notes

1. Here, I follow the anthropological notion of universalism which opposes a universalism that posits an absolute and shared human essence to a relativism that affirms the diversity of cultural identities (see Lamont and Aksartova, 2002, p. 19).
2. See Lamont and Molnár (2002b) for a comprehensive account of the use of the concept of boundary in the social sciences.
3. The history of the frontier (*frontière*) goes progressively and irreversibly from the empty zone of separation to the linear boundary (*limite*). The linear frontier was a rarity until modern periods because of the virtual absence of continuous lines of fortification. See Nordman (1998) and Anderson (1996).
4. Nordman (1998, p. 33) identifies two kinds of divides or limits (*bornes*) to indicate the existence of a boundary: natural divides include rivers, a forest, a mountain range or other topographic formations, whilst artificial divides refer, for example, to walls, a fortress, or a trench.
5. Translations of subtitles in the Portuguese language concerning dialogue in *The Suspended Step of the Stork* are the present author's work.
6. Territorial expansion in Greece led finally to the annexation of western Thrace and Greek Macedonia after the Balkan Wars.
7. The novella by Álvaro Cunhal, the historical leader of the Portuguese Communist Party, was found in 1974 in the aftermath of the 25 April Revolution which ended the authoritarian Estado Novo regime. *Cinco Dias, Cinco Noites/Five Days, Five Nights* (1994) was later published under the pseudonym of Manuel Tiago.
8. Translations of dialogue in the Portuguese language in *Five Days, Five Nights* are the present author's work.
9. The way people imagine their relation with both local and national dimensions of space can be described by the idea of a 'mental map'.

References

Alexander, J. C. (2003) *The Meanings of Social Life – A Cultural Sociology*, New York: Oxford University Press.

Anderson, M. (1996) *Frontiers - Territory and State Formation in the Modern World*. Cambridge: Polity Press.

Barth, F. (1969) *Ethnic Groups and Boundaries*. Oslo: Norwegian University Press.

Barth, F. (1994) 'Enduring and Emerging Issues in the Analysis of Ethnicity', in H. Vermuelen and C. Govers (eds) *The Anthropology of Ethnicity: beyond 'Ethnic Groups and Boundaries*, Amsterdam: Het Spinhuis.

Baud, M. and Schendel, W. V. (1997) 'Toward a Comparative History of Borderlands', *Journal of World History* 8(2): 211–242.

Bordwell, D. (1997) *On the History of Film Style*. Cambridge: Harvard University Press.

Castles, S. (2003) 'Towards a Sociology of Forced Migration and Social Transformation', *Sociology* 37(1): 13–34.

Chard, C. (1999) *Pleasure and Guilt on the Grand Tour - Travel Writing and Imaginative Geography 1600–1830*, Manchester: Manchester University Press.

Delanty, G. (1995) *Inventing Europe – Idea, Identity, Reality*. Houndmills, London: Macmillan Press.

Delanty, G. (1996) 'The Frontier and Identities of Exclusion in European History', *History of European Ideas* 22(2): 93–103.

Douglas, M. (1966) *Purity and Danger*. London: Routledge and Kegan Paul.

Eisenstadt, S. N. (2001) 'The Civilizational Dimension of Modernity: Modernity as a Distinct Civilization', *International Sociology* 16(3): 320–340.

Entrikin, J. N. (1999) 'Political Community, Identity and Cosmopolitan Place', *International Sociology* 14(3): 269–282.

Ferrão, J. (2002) 'Portugal, Três Geografias em Recombinação: Espacialidades, Mapas Cognitivos e Identidades Territoriais', *Lusotopie* 2: 151–158.

Fortuna, C. (1997) 'Destradicionalização e Imagem da Cidade - o Caso de Évora', in C. Fortuna (ed.) *Cidade, Cultura e Globalização*, Oeiras: Celta.

Hartshorne, R. (1938) 'A Survey of the Boundary Problems of Europe', in C. C. Colby (ed.) *Geographic Aspects of International Relations*, Chicago, Illinois: The University of Chicago Press.

Horton, A. (1997) *The Films of Theo Angelopoulos – A Cinema of Contemplation*. Princeton, New Jersey: Princeton University Press.

Kelly, L. (2003) 'Bosnian Refugees in Britain: Questioning Community', *Sociology* 37(1): 35–50.

Lamont, M. and Aksartova, S. (2002) 'Ordinary Cosmopolitanisms – Strategies for Bridging Racial Boundaries Amongst Working-Class Men', *Theory, Culture & Society*, 19 (4): 1–25.

Lamont, M. and V. Mólnar (2002) 'The Study of Boundaries in the Social Sciences', *Annual Review of Sociology* 28: 167–195.

Lamont, M. and Thévenot, L. (eds) (2000) *Rethinking Comparative Cultural Sociology: Repertoires of Evaluation in France and the United States*, Cambridge: Cambridge University Press.

Lazaridis, G. (1996) 'Immigration to Greece: a Critical Evaluation of Greek Policy', *New Community* 22(2): 335–348.

Leed, E. (1991) *The Mind of the Traveller: from Gilgamesh to Global Tourist*. Basic Books.

Llobera, J. R. (1994) 'Anthropological Approaches to the Study of Nationalism in Europe. The Work of Van Gennep and Mauss', in V. A. Goddard, J. R. Llobera and C. Shore (eds) *The Anthropology of Europe – Identity and Boundaries in Conflict*, Oxford, Providence, USA: Berg.

Llobera, J. (2003) 'The Concept of Europe as an Idée-force', *Critique of Anthropology*, 23(2): 155–174.

Martins, H. (1971) 'Portugal', in M. Archer and S. Giner (eds) *Contemporary Europe: Class, Status and Power*. London: Weidenfeld and Nicolson.

Matos, J. (1999) *Pelos Espaços da Pós-Modernidade – A Literatura de Viagens Inglesa da Segunda Guerra à Década de Noventa*. Porto: Edições Afrontamento.

Monteiro, N. G. and Pinto, A. C. (1998) 'Cultural Myths and Portuguese National Identity', in A. C. Pinto (ed.) *Modern Portugal*. Palo Alto: The Society for the Promotion of Science and Scholarship.

Moretti, F. (2000) *The Way of the World: The Bildungsroman in European Culture*. London and New York: Verso.

Morin, E. (1990) *Penser l'Europe*. Paris: Gallimard.

Morley, D. and Robins, K. (1995) *Spaces of Identity – Global Media, Electronic Landscapes and Cultural Boundaries*. London: Routledge.

Nordman, D. (1998) *Frontières de France – De l'Espace au Territoire XVI–XIX Siècle*. Éditions Gallimard.

O'Dowd, L. (2001) 'State Borders, Border Regions and the Construction of European Identity', in M. Kohli and M. Novak (eds) *Will Europe Work? Integration, Employment and the Social Order*. London: Routledge.

Peckham, R. (2000) 'Map Mania: Nationalism and the Politics of Place in Greece, 1870–1922', *Political Geography* 19(1): 77–95.

Peixoto, P. (2002) 'Os Meios Rurais e a *Descoberta* do Património', Coimbra: Oficina do CES.

Pierron, M.-J. (1995) 'La Ville-frontière: de Wenders à Angelopoulos', *CinémaAction*: 138–143.

Porter, D. (1991) *Haunted Journeys - Desire and Transgression in European Travel Writing*, Princeton, New Jersey: Princeton University Press.

Rebelo, J. (1998) *Formas de Legitimação do Poder no Salazarismo*, Lisboa: Livros e Leituras.

Rosas, F. (1998) 'Salazarism and Economic Development in the 1930s and 1940s: Industrialization Without Agrarian Reform', pp. 88–101 in A. C. Pinto (ed.) *Modern Portugal*. Palo Alto: The Society for the Promotion of Science and Scholarship.

Roudometof, V. (2001) *Nationalism, Globalization, and Orthodoxy – The Social Origins of Ethnic Conflict in the Balkans*. Westport and London: Greenwood Press.

Rovisco, M. (2003) *European Films of Voyage: Nation, Boundaries and Identity*, Ph.D. dissertation in Sociology, York: University of York.

Said, E. (1993) *Culture and Imperialism*. London: Vintage.

Santos, B. S. (1994) *Pela Mão de Alice – O Social e o Político na Pós-Modernidade*. Porto: Edições Afrontamento.

Sassateli, M. (2002) 'Imagined Europe – The Shaping of a European Cultural Identity through EU Cultural Policy', *European Journal of Social Theory* 5(4): 435–451.

Schlesinger, P. (1991) *Media, State and Nation: Political Violence and Collective Identities*. London: Sage.

Simmel, G. (1970) 'The Stranger', pp. 143–149 in D. D. Levine (ed.) *On Individuality and Social Forces: Selected Writings*. Chicago: University of Chicago Press.

Strassoldo, R. (1982) 'Boundaries in Sociological Theory: A Reassessment', in R. Strassoldo and G. D. Zotti (eds) *Cooperation and Conflict in Border Areas*, Milano: Franco Angeli.

Stråth, B. (2002) 'A European Identity – To the Historical Limits of a Concept', *European Journal of Social Theory* 5(4): 397–401.

Tiago, M. ([1994] 2001) *Cinco Dias, Cinco Noites*. Lisboa: Editorial «Avante!».

Vandenberghe, F. (2005) *As Sociologias de Georg Simmel*. Belém: EDUFPA.

Villaverde Cabral, M. (1992) 'Portugal e a Europa: Diferenças e Semelhanças', *Análise Social* XXVII (118–119): 943–954.

Wertsch, J. (2002) *Voices of Collective Remembering*. Cambridge: Cambridge University Press.

Williams, C. and Smith, A. (1983) 'The National Construction of Social Space', *Progress in Human Geography* 7(4): 502–518.

Zolberg, A. and Long, L. (1999) 'Why Islam is Like Spanish: Cultural Incorporation in Europe and the United States', *Politics & Society* 27(1): 5–3.

THE MEDIA AND THE SYMBOLIC GEOGRAPHIES OF EUROPE: THE CASE OF YUGOSLAVIA

Sabina Mihelj

Introduction

Following the dramatic reconfiguration of the map of Europe after 1989, scholars dedicated considerable attention to questioning the seemingly neutral categories used to define the borders of Europe and European regions and cultures. 'Balkan culture' – just as 'European', 'western' or 'eastern' culture – was no longer conceived purely in terms of objective criteria, but also as 'an argument over meanings and definitions, advanced by particular people, in particular places, for particular purposes [...] and reconfigured in response to changing social, cultural and political processes' (Bracewell and Drace-Francis, 1999, p. 56).

Often inspired by Edward Said's influential study of the western conceptions of the Orient (1978) – but for the most part also well aware of the limitations of applying Said's approach to the Balkans – historians, anthropologists and political scientists have produced a number of detailed studies exploring the historical formation and variegated uses of western perceptions of the Balkans, eastern Europe, and other regions designated as 'eastern' (see Corn, 1991, 2003; Allcock and Young, 1991; Wolff, 1994, 2001; Todorova, 1997; Gingrich, 1998; Goldsworthy, 1998; Norris, 1999; Neumann, 1999; Jezernik, 2004). Taken-for-granted mental mappings were dismantled as human constructions, developed only in the modern era; the habit of seeing Europe as divided into eastern and western Europe became a commonplace in the Enlightenment period, replacing the earlier division of Europe into the South and the

North (see Wolff, 1994), while the discourse of Balkanism as a place of disorder, decay, underdevelopment, violence and totalitarian measures was formed over the course of two centuries and crystallized with the Balkan wars and World War I (see Todorova, 1997).

It has been repeatedly asserted that the choice of criteria used to define borders and differences between regions such as eastern and western Europe, or Central Europe and the Balkans, is never neutral. Instead, it is guided by particular ideologies and supported by (and supportive of) specific relationships of domination and subordination in the international political arena. For example, the various historical definitions of the Balkans were criticized for their inclination to represent the Balkans in terms of their similarity to or deviation from Europe, associating them with backwardness and barbarity and subordinating them to the developed and civilized Europe, thus justifying the existing relationships between western European and Balkan states in a particular period (see Todorova, 1997, Bracewell and Drace-Francis, 1999).

While the majority of these studies have focused almost exclusively on the ways western European authors imagine the Balkans and the East, some scholars have pointed to the fact that these symbolic geographies are far from being an exclusively western European product. Instead, they are often (re-)produced locally, within the despised regions themselves, whose inhabitants tend to internalize the categories applied to them by their western observers, or use them to distinguish themselves from the inhabitants of neighbouring states (Bakić-Hayden and Hayden, 1992, Bakić-Hayden, 1995; Todorova, 1997; Norris, 1999; Goldsworthy, 1999; Iordanova, 2000, 2001; Bjelić and Savić, 2002). The symbolic geographies gaining ground in socialist Yugoslavia in the late 1980s and proliferating in the successor-states after its disintegration are a case in point. In their pioneering essay, Milica Bakić Hayden and Robert M. Hayden highlight several statements of Slovenian and Croatian public figures to demonstrate how they were involved in constructing a chain of 'nesting Orientalisms', claiming 'a privileged "European" status for some groups in the country while condemning others as "Balkan" or "Byzantine", hence non-European and Other' (Bakić Hayden and Hayden, 1992, p. 5). Following their arguments, several scholars have set out to explore the appropriations and uses of the East/West and Europe/Balkans dichotomies among peoples once sharing a common Yugoslav state, mostly focusing on Croatia or Serbia (Buden, 2000, 2000; Čolović, 1997; Rihtman-Auguštin, 2000 [1997]; Guzina, 1999; Jansen, 2001a; Živković, 2001; Šakaja, 2001, 2003; Czerwiński, 2003; Lindstrom and Razsa, 2004; Volčič, 2005) and, to a lesser extent, Slovenia (Patterson, 2003, Mihelj, 2004a) and Bosnia-Herzegovina (Helms, 2004).

When looking for examples of local appropriations of various symbolic geographies in the territory of former Yugoslavia, authors often turn to the media – either to the mass media such as the daily and weekly press or television and radio news programmes, or to media with somewhat more limited audiences such as cultural and scientific journals, collections of essays, travelogues and memoirs. However, they virtually never reflect on the characteristics of sources used, for example their range of distribution and audiences, and on what this can reveal about the potential impact of particular appropriations of symbolic geographies within the larger society. For example, the fact that a specific mental mapping is used or discussed in a cultural or scientific journal cannot be seen as a confirmation of the thesis that this particular mapping holds a hegemonic position in the whole society. Furthermore, even if one finds firm evidence of a widespread use of a particular symbolic mapping in the mainstream mass media, this does not provide many clues about whether and how this mapping is used and appropriated among the wider population on a day-to-day basis. In existing examinations of symbolic geographies in Yugoslavia, such considerations are almost non-existent, and investigations of mental maps shared by the wider population are extremely rare (see Šakaja, 2001 for an exception). For the most part, authors tend to draw strong conclusions about the hegemonic status of particular mental mappings without ever expressing doubts about the representativeness of their sources and samples. Lately, some authors have explicitly argued against the tendency to treat Balkanist and similar discourses as all-pervasive, persistent and uniform, and called for more attention to be paid to 'those countervailing forces that may curb the power of rhetoric with aspirations to hegemony' (Patterson, 2003, p. 141). If this is to be achieved, more thought will have to be given to the selection of sources, and to those characteristics of sources that may affect the formation, proliferation, reification or negotiation of particular symbolic geographies.

In an attempt to fill a part of this gap, this chapter first provides a brief summary of existing investigations of symbolic geographies in Yugoslav republics (and later independent states), identifying frequently discussed aspects as well as those that have not yet received enough attention. Then it addresses the question of whether, how, and to what extent particular media have supported the spreading of symbolic geographies analyzed in existing studies. The chapter does this by discussing the institutional arrangements of the Yugoslav mass media (press, radio and television), looking at how they might have contributed to the success of these symbolic geographies, but also how they may have countered the (re)production of such symbolic geographies, and were overlooked in existing studies. Then it examines a range of 'media' in a wider sense of the word that have participated in the formulation, appropriation and negotiation of various symbolic geographies. After investigating cultural journals and films – the media that have already been considered as sites of (re)production and negotiation of symbolic geographies – it concludes by considering means of communication whose role in the formulation and appropriation of symbolic geographies has not yet been fully acknowledged: music, food, and street names.

An overview of existing studies of symbolic geographies of Yugoslavia and successor states, 1980s–1990s: main issues and omissions

In the course of the disintegration of Yugoslavia and the subsequent wars, the existing network of collective identifications, including collective memories and collective attachments to space, changed substantially. Before the 1980s, the symbolic divisions of Yugoslavia into its western and eastern (or, even more often, northern and southern) parts were not seen as insurmountable. Each republic and province was seen as an integral part of Yugoslavia, and thus as a component of a compact spatial and cultural unit that, albeit internally diverse, actually shared a recognizable common identity: it was neither western nor eastern. The positioning of Yugoslavia as neither-western-nor-eastern penetrated into various levels of public discourses in the federation, and various spheres of Yugoslav culture, politics and society were envisaged in terms of their proximity to and distance from both East and West. Particularly with the formation of the Non-Aligned Movement in the 1950s, the neither-eastern-nor-western positioning became one of the central elements of the socialist Yugoslav political identity and one of the defining traits of the Yugoslav foreign policy (Petković, 1985). Furthermore, the distinguishing feature of the socialist Yugoslav economy (in the official jargon: self-management), in which business was run in the form of cooperatives, was thought of as combining elements of both Soviet (that is, eastern) economic statism as well as the western free market economy (see Sirc, 1979). Finally, even the descriptions of the Yugoslav media were sometimes framed in terms of their neither-eastern-nor-western character. Pedro Ramet, for example, opened his chapter on the socialist Yugoslav press with the following assertion: 'Like Yugoslavia itself, the Yugoslav press is neither of the East nor of the West' (Ramet, 1985b, p. 100).

In the late 1980s, this particular positioning, coupled with the image of Yugoslavia as characterized by unity-in-diversity, was gradually disappearing. Rather than being a source of pride, the internal diversity of Yugoslavia was, among political elites in Slovenia and Croatia, increasingly seen as an obstacle to its further development, and Yugoslavia itself begun to be perceived as an unviable mixture of incompatible civilizations or cultures: the western and the eastern one, the European and the Balkan one. Milica Bakić-Hayden has summarized this change at the level of symbolic geography and its impact in the following way: As a political entity, the former Yugoslavia encompassed traditional dichotomies such as East/West and their nesting variants (Europe/Asia, Europe/Balkans, Christian/Muslim), largely neutralizing their usual valorisation. With the destruction of this neutralizing framework, the revalorisation of these categories, now oppositions rather than simply differences, has resulted in the destruction of the living communities that transcended them (Bakić-Hayden, 1995, p. 927). However, this transformation did not occur in the form of a linear, unequivocal development, nor was it ever complete. Rather, as becomes evident from a careful consideration of examples quoted in various studies of symbolic geographies in the region, a range of distinct labels and dichotomies was used to draw a number of differently positioned symbolic borders; these labels, dichotomies and borders were invested with a wide array of meanings; and finally, they were not only used to draw distinctions *between*, but also *within* particular (formerly) Yugoslav nations. Without aiming to be exhaustive, these variations can be divided into four main groups:

(a) *geographical labels and dichotomies*: East vs. West, eastern Europe vs. western Europe, Europe vs. the Balkans, Europe vs. south-eastern Europe; Europe vs. Asia, Central Europe vs. the Balkans, Central Europe vs. eastern Europe, Mediterranean vs. the Balkans etc.
(b) *political borders seen as dividing lines between Europe and the Balkans, West and East etc.*: the border between Slovenia and Croatia, the border between Croatia and Serbia, the border between Croatia and Bosnia and Herzegovina, the border between Bosnia and Herzegovina and Serbia and Montenegro, the border between Vojvodina and the rest of Serbia, the border between Serbia and Bulgaria, Romania and Albania
(c) *intra-national divisions seen as running in parallel to symbolic geographical divisions*: the rural/urban divide, divisions between different political options within the same state

(d) *meanings and values attached to geographical labels*: Europe as a symbol of civilization vs. Europe as a symbol of decadence, Balkans as a symbol of barbarism vs. Balkans as a symbol of vitality, Europe as a union of nations vs. Europe as a union of states

In existing studies, by far the most often discussed mappings are those employing the dichotomies Europe vs. the Balkans and Central Europe vs. the Balkans. Several authors have argued that in particular Slovenia and Croatia have reserved the labels Europe or Central Europe for themselves, disparaging the rest of the former Yugoslavia as Balkan (Guzina, 1999; Buden, 2000; Rihtman-Auguštin, 2000 [1997]; Lindstrom and Razsa, 2004). While most studies have focused exclusively on examples where the symbolic division is drawn in an unambiguous manner, some attempts have recently been made to identify cases where such mappings were disputed; for example, it was shown that several Slovenian intellectuals disputed the labelling of Slovenia as Central European (Patterson, 2003, pp. 119–121), and that in the months preceding the plebiscite for Slovenian independence in 1990, there was some evidence of newspaper articles explicitly countering the Balkanist discourse (Mihelj, 2004b, p. 214). Other symbolic geographical labels have received far less attention, although there is reason to suspect that some of them – in particular the North vs. South division – were a common element of everyday conversations, and were not, like the discussions about Central Europe, Europe and the Balkans, limited to intellectual and political elites. As Zlatko Skrbiš (1999) argued, the notion of the 'Southerner' was used extensively in Slovenia to refer to people from other former Yugoslav republics, and the South 'has been commonly perceived in a symbolic fashion as the personification of economic underdevelopment, hot-bloodedness and, most often, otherness' (Skrbiš, 1999, p. 121).

The shifting nature of symbolic borders is another often-raised issue in existing examinations of symbolic geographies in Yugoslavia. As several authors argued, the opinion-leaders in virtually each Yugoslav republic and province aimed to portray themselves and their co-nationals as authentically European, and presented the federal unit located towards the East or South as comparably less European and more Balkan. The often-quoted ironic excerpt from Slavoj Žižek's *The Fragile Absolute* captures the logic of this chain-reaction very well:

> For the Serbs, [the Balkans] begin down there, in Kosovo or in Bosnia, and they defend the Christian civilisation against this Europe's Other; for the Croats, they begin in orthodox, despotic and Byzantine Serbia, against which Croatia safeguards Western democratic values; for Slovenes it begins in Croatia and we are the last bulwark of the peaceful *Mitteleuropa*; for many Italians and Austrians they begin in Slovenia, the Western outpost of the Slavic hordes; for many Germans, Austria itself, because of its historical links, is already tainted with Balkan corruption and insufficiency; for many North Germans, Bavaria, with its Catholic provincial flair, is not free of a Balkan contamination...(Žižek, 2000, pp. 3–4)

Such uses of Europe and the Balkans undeniably present yet another extension of the dark side of the history of the idea of Europe: a history of division and exclusion rather than unity and inclusion (see Delanty, 1995). But however amusing, accounts such as Žižek's fail to acknowledge that within individual states symbolic geographies were used in a number of different ways, and were invested with a wide range of meanings and values. For example, divisions such as Europe vs. the Balkans would often be applied to divisions *within* particular nations, thus carving them into a European part and a Balkan part. Frequently, such a use of this symbolic geography would coincide with the rural/urban divide in the country (see Rihtman-Auguštin, 2000 [1997], pp. 227–8; Jansen, 2001b, p. 51; Norris, 1999, p. 163; Helms, 2004, pp. 12–14; Volčič, 2005), while in other cases, the label 'Balkan' would be applied to political enemies, be it nationalists or anti-nationalists (see Rihtman-Auguštin, 2000 [1997], pp. 218–9 Jansen, 2001a, pp. 7–9), including Croatian communists such as Tito (Jansen, 2001a, p. 6).

Another aspect that has not yet received enough attention is the variation in the meanings and values attached to different geographical labels within the same states. Most studies remain limited to the examination of the uses of Europe as a symbol of everything valuable and desirable, and as a chief instrument of exclusion. Milica Bakić-Hayden's approach is a case in point; according to her, 'the symbolic power of "Europe" to represent "civilized", "enlightened" or "progressive" in Yugoslav debates created a standard against which peripheral European countries could judge their multiple selves in competition against each other' (Bakić-Hayden, 1995, p. 927). Although it is true that the idea of Europe 'has become a political football by which groups can distinguish themselves from others' (Delanty, 1995, p. 135), this has not been the only use of Europe (and associated dichotomies such as Europe vs. Balkans) within Yugoslavia. The symbolic geographies circulating in Serbia and Bosnia-Herzegovina in the 1990s are a case in point; although Serbs can, and sometimes do, present themselves as European and/or transfer the stigma of being Balkan to their neighbours, they far more often employ a rhetorical strategy of self-balkanization or self-exoticization, applying the Balkan stigma to themselves, and thus agreeing that they are wild, uncivilized, primitive, violent, irrational, passionate etc. However, in such cases, the characteristics associated with the Balkans are often charged with a positive value: the wildness, barbarity, passions, primitiveness etc. are presented as precious features which the peaceful, civilized, cold-blooded and decadent Europe misses (Živković, 2001, pp. 12–15; Jansen, 2001a, pp. 12–14; Helms, 2004, pp. 13; Volčič, 2005, pp. 164).

This duality in the symbolic portrayal of the Balkans is, as a rule, matched by a duality in the symbolic portrayal of Europe. For example, Ivan Čolović (1997, p. 43) argued that in Serbia, 'the mythical image of the West as the embodiment of justice, culture and prosperity, within which the Balkans and Serbia in the Balkans acquire the infamous role of the symbol of underdevelopment, primitivism and barbarity' was paralleled by the resurgence of the belief that contemporary Europe has succumbed to the temptations of humanism and materialism and that, in fact, the Serbs are the only true heirs and guardians of the authentic European values (see Volčič 2005 for a detailed examination of such understandings among young Serbian intellectuals). In Croatia, too, Europe would quite often be portrayed in overtly negative tones; especially after it became obvious that western Europe did not intend to enter the Yugoslav wars as Croatia's ally, Europe ceased to function as an unambiguously positive symbol. Similarly, as in Serbia, Europe was believed to be corrupted by materialism and consumerism, while Croatia was thought of as a true hero and a martyr, a country which is yet again sacrificing itself for Europe, in order to defend its true values (Christianity) from the onslaught of the East (Islam) (Buden, 2000). Finally, such resentment towards Europe would occasionally surface in Slovenia as well. For example, in the late spring of 1992, when Slovenia had to cope with increasing numbers of war refugees from Bosnia, and western countries were unwilling to intervene, Europe was represented as extremely irresponsible and accused of betraying its own values and rules (Mihelj, 2004b, pp. 297–302).

Finally, the notion of Europe as a union of nations, or more precisely nationally homogeneous states, can also be identified in existing discourse. Within Slovenia and Croatia, the disintegration of multinational Yugoslavia and the creation of nationally homogeneous states was often presented as an unavoidable part of becoming European, and thus – given the symbolic values usually attached to Europe – of becoming civilized and modern. Precisely at the moment when Europe was striving to transcend the nation state model, the political elites in Slovenia and Croatia have come to equate Europeanization with the formation of nationally homogeneous states (see Mihelj, 2004b, pp. 208–9). Obviously, such an understanding of Europe required the elimination of Yugoslavia and all the layers of identification and attachment that were not compatible with the nation state model. The new supranational collective attachment to Europe could only be established if the previous supranational collective attachment to Yugoslavia was abolished, and the state structure supporting it disintegrated.

The mass media institutions in socialist Yugoslavia: decentralisation as a guarantee of democracy or a prelude to disintegration?

Arguably, the institutional arrangement of the mass media in Yugoslavia rendered the media very open to embracing the symbolic geography that divided Yugoslavia into incompatible geo-cultural zones and, ultimately, into individual nation states. At least two major reasons can be identified: (a) the fact that the Yugoslav media were conceived primarily as crucial propaganda and educational institutions supporting the views of the governing elites, and not as public forums where different opinions could be openly confronted, and (b) the fact that the Yugoslav media system was heavily decentralized and included only a very limited number of pan-Yugoslav media.

The mass media in socialist Yugoslavia were – in tune with what came to be known as 'the Leninist theory of the press' – conceived primarily as propaganda and educational institutions, and not as public forums where different opinions could be openly confronted. The constitutionally guaranteed 'freedom of the press and other media of information and public expression' had clear limits (see Thompson, 1999, pp. 8–10), and combined with a high level of self-censorship, resulted in a culture inside of which journalists were used to being the loud-speakers of political authorities; once these started employing the new symbolic geography, most of the mass media followed suit.

One certainly needs to acknowledge that the Yugoslav media were more abundant, varied and unconstrained than in any other state in the eastern bloc (Thompson, 1994, p 5), and that in particular the Yugoslav press has repeatedly resisted or circumvented political controls, endeavouring to 'expand the sphere of its legitimate activity, even to the point of criticizing the government itself' (Ramet, 1985b, p. 101). However, due to the particular way in which the media system was decentralized, a large proportion of these criticisms took the form of a conflict between a particular Yugoslav republic or province (or its political elite) and the federal government, and for the most part, different views were not confronted in a single public sphere, but coexisted in different, separated public spheres.

During the 1960s and 1970s, the political power in Yugoslavia was devolved from the central (federal) organs to the six republics (Bosnia-Herzegovina, Croatia, Macedonia, Montenegro, Serbia and Slovenia), two autonomous provinces within Serbia (Kosovo and Vojvodina), and their respective branches of the League of Communists of Yugoslavia. At the same time, the media were also decentralized: each republic and province had its own major daily newspaper, its own radio and television centre, as well as several regional newspapers, radio stations and weekly magazines, which effectively functioned as forums for the discussion of views held by respective republican or provincial elites. Only a small number of the media, with very limited audiences, would address the population of the federation as a whole. While it is true that such an organization of the media system led to a situation where, as Pedro Ramet (1985b, p. 102) argued, 'diametrically opposed viewpoints [could] be found in the daily papers of the different republics on a recurrent basis', one cannot agree with the claim that such a federalization of the press 'contributes to the openness by undermining central direction and by getting people accustomed to seeing different interpretations voiced in the press'. The decentralization of the Yugoslav media system has indeed provided opportunities for the constitution of a number of distinct and differently oriented public forums, yet it needs to be kept in mind that these forums were not equally accessible to the whole of the Yugoslav population, and did not provide an opportunity for different views to be read, listened to or discussed by the same population. With the exception of a couple of pan-Yugoslav media, the mass media in Yugoslavia were controlled at the republican or provincial level, increasingly geared for republican or provincial audiences, and distributed and consumed within the particular republic or province.

Furthermore, in most republics, especially the ethnically most homogeneous ones, these audiences would to a large extent correspond to particular Yugoslav nations. To put it differently: various Yugoslav nations had come to constitute exclusive imagined communities of the various republican/provincial media long before there were any visible signs of the disintegration of Yugoslavia (Vogrinc, 1996, p. 13), and the banal nationalism (Billig, 1995) had long been built into the repetitive forms of press, radio and television in particular republics. Keeping this in mind, Gertrude J. Robinson's (1997, pp. 192–99) interpretation of the diversity of opinions in the Yugoslav media seems to be more correct than the one given by Ramet: 'In this country's multinational setting, content is selected with ethnic priorities in mind; this tends to encourage regionalism and fosters hermetic points...These could be potentially destructive of federal unity by undermining the search for and definition of mutually acceptable political alternatives' (Robinson, 1977).

One could hardly ignore the role of existing institutional arrangements of the mass media in socialist Yugoslavia for the spread of symbolic geographies that went against the notion of a common Yugoslav geo-cultural space. However, such uses were not inevitable, and moreover, not all the mass media were equally open to such uses. Firstly, the overlapping of republican mass media audiences and individual Yugoslav nations was far from perfect. Particularly in Bosnia and Herzegovina, where no clear national majority was in place, it was virtually impossible to address the whole audience in nationalist terms, as a homogeneous nation. Furthermore, regardless of the ethnic composition, not all the mass media in particular republics would be equally prone to endorse the new, exclusivist symbolic geography; at least some media would, as a rule, subvert such geographies or support different ones. And finally, besides the republican mass media, which were more prone to endorsing the new symbolic geography, the Yugoslav media landscape also included a number of pan-Yugoslav media: the daily *Borba* [*Struggle*], the news agency *Tanjug*, *Radio Yugoslavia* and the short-lived TV station *Yutel*. These media could address the members of various nations as Yugoslavs and could foster their attachment to a common, Yugoslav space. One could thus expect these media to adopt an alternative symbolic geography of Yugoslavia, one which envisaged Yugoslavia as a whole as a future member of the European community of states.

Still, one should also acknowledge that these pan-Yugoslav media were rare and did not attract a wide audience. Thus, most of information issued by federal bodies, which largely countered the disintegration and supported a vision of Europe that included the whole of Yugoslavia, did not reach the audiences through the federal media, but through republican ones, who would regularly criticize and delegitimize ideas fostered by the federal bodies (see Mihelj, 2004b, pp. 221–2 for the case of Slovenia). The fate of *Yutel*, the only pan-Yugoslav TV station, established in 1990, is a case in point. Due to severe budgetary limitations, *Yutel* had to rely on the technical as well as political support of television centres in other republics and provinces. Most of the time, the majority of republican and provincial TV centres declined *Yutel*'s offer to broadcast its programmes on any of its channels, and even when they accepted, they tended to broadcast *Yutel*'s programmes outside of prime-time, and on channels whose signals did not cover the whole territory of a particular republic (Thompson, 1994, pp. 38–49). Moreover, *Yutel*'s news items were regularly criticized and ridiculed by republican TV news; they claimed *Yutel*'s reports were being biased in favour of another republic's point of view (see Kurspahić, 2003, p. 69). Arguably, the re-framing of *Yutel*'s reporting through the particular republican mass media, in combination with *Yutel*'s limited coverage, importantly diminished the already limited influence of *Yutel*'s narratives, geographies, and collective identifications. Therefore, its representations of Europe and Yugoslavia and the concomitant collective attachments, just as representations produced by the other federal media, could hardly rival those offered by the mainstream republican or regional media in each particular republic or regions. Furthermore, as most federal media (with the exception of *Yutel*) were based in Belgrade, they were, from 1987 onwards, exposed to the pressures of Milošević's regime, and after a period of resistance, mostly succumbed to them (see Nenadović, [1996] 2000). Finally, the reach of collective representations offered by the various independent opposition media inside individual republics – such as *Radio B92* and the TV station *Studio B* in Belgrade or *Radio 101* in Zagreb – was also very limited. Although representations provided by these media should not be ignored – since they may show that nationalist, Balkanist and other discourses were not as homogeneous and all-pervasive as is sometimes suggested – it also has to be admitted that until political control exerted over them was eased, they could not seriously rival the mainstream ones.

A wider conception of the 'media'

Radio, television and the periodical press provided the main forums within which the new symbolic geography was negotiated, solidified and reified as a regular frame of political and intellectual discourse. However, the gradual formulation of the new symbolic geography, its inculcation into everyday practices, as well as its various appropriations and negotiations – including explicit rejections – often took place outside of the mass media. In order to gain insight into these aspects of the development and spread of the new symbolic geography, one needs to take into account a whole range of media in the wider sense of the word: fiction books, theatre plays, collections of poems, cultural journals, music, street names, monuments, stamps, state symbols, food, and so on. Each of these media played a particular role in the process – some of them being crucial for the initial formulation of the new symbolic geography, others for its final reification.

Although this thesis would need additional empirical investigation, there is enough evidence to suggest that the media produced by intellectuals – literary works, theatre plays as well as essays published in cultural journals – were among those media where the new symbolic geography of Yugoslavia was first formulated. Several authors have argued that it was in the field of cultural production that the legitimacy of the socialist Yugoslavia was first shaken, thus preparing the grounds for similar changes to take place in other fields, including politics. According to Pedro Ramet (1985a), a veritable 'apocalypse culture' developed in Yugoslavia in the 1980s. Associated with 'pessimism, gloom, resignation, escapism of various kinds, and a feverish creativity', as well as 'normlessness and anomie', this culture was characterized by an 'openness to radically new formulas', springing from the sense that the Yugoslav system 'has arrived at a historical turning point, that it is, so to speak, at the "end of time"' (Ramet, 1985a, p. 3). Contributors to this apocalypse culture – writers, poets, theatre directors, intellectuals, even politicians – saw themselves as social critics, voicing criticisms of the system, or as prophets, warning of dangers ahead and offering new visions of the future. Through various forms of cultural production, a mixture of different diagnoses and prophetic visions was offered, ranging from fiercely nationalist to entirely a-nationalist ones, and some of them included elements of the divisive symbolic geography that was to become so widespread in the years to come.

From the mid-1980s, poets, writers and intellectuals in Slovenia have been discussing Slovenia's links with Central Europe in various cultural journals (*Naša sodobnost* [Our Contemporary Times], *Naši razgledi* [Our Views], *Primorska srecanja* [Encounters of the Littoral] and others), and organizing the annual literary festival Vilenica in order to put Slovenia on the cultural map of Central Europe. Finally, the cultural journal *Nova revija* [*New Review*] published the *Contributions for the Slovenian National Program* (1987), a collection of essays written by writers, poets and other intellectuals professing an independent Slovenia, most of them underpinned by a clearly nationalist mind-set (Mihelj, 2004b, pp. 162–7). The journal also provided many examples of the harshest variant of the Slovenian derogatory attitudes towards the Balkans (Patterson, 2003, p. 118). In Serbia, a similar role in the crystallization of the new geopolitical positioning of Serbia was played by among others, *Književne novine* [*Literary Newspaper*] and *Duga*. However, as Patterson (2003, p. 119) rightly warns, one should beware of assuming that the symbolic geographies used in these journals were monolithic; even in *Nova revija*, Balkanism was not a uniform and all-pervasive strategy. Furthermore, it may well be that other cultural or scientific journals and similar outlets – for example, those which did not wholeheartedly embrace nationalism and virtually never appear among sources used in existing studies of symbolic geographies in the region – were entirely devoid of such Balkanist mappings, or perhaps even openly critical of them.

Besides the media and cultural products created by and aiming at highly educated and profiled audiences, other, more popular or more widely consumed forms of cultural production can also be considered as a site where the old spatial attachments (along with collective memories and collective identifications in the wider sense) were being challenged and renegotiated: music, films, TV shows and series, street names and monuments, state symbols, bank notes, stamps, and even restaurant menus. Interestingly, it is precisely those authors who have examined popular cultural forms who have most often pointed out that the stigma of the Balkans is not always transferred to the neighbours, and that the label 'Balkan' is not necessarily associated with negative characteristics. Instead, it is sometimes consciously and explicitly accepted as a valid self-designation and even associated with positive values and represented as an object of desire. Unfortunately, out of all the diverse popular cultural forms, only film has received a fair measure systematic treatment in existing literature on the topic. According to Dina Iordanova (2001), the preferred mode of self-expression from many Balkan film-makers producing their films in the 1990s was a specific voluntary 'self-exoticism': 'the "otherness" of the Balkans, which may have originated in the West, is gradually taken up and internalised by local directors who claim to represent the Balkans "from within"'. One of the traits of films by Balkan film-makers where such self-exoticisation becomes clearly evident is, to her view, the narrative structure. According to her analysis, a number of films are structured around the same plot, drawing on the travelogue tradition and on the figure of the visiting western protagonist, depicting 'well-balanced and presumably sane Westerners who venture into the Balkan realm of barbarity' (Iordanova, 2001, p. 61).

Other popular, widely consumed cultural forms have received only cursory treatment in existing studies of symbolic geographies in the region, but promise to be a rich source; one of them is music. As Alexander Kiossev argued, both in the territory of former Yugoslavia as well as in other Balkan states, the 1990s saw an unprecedented rise in the popularity of certain types of rock and folk music (or a mixture of both), which successfully replaced English and American music in clubs and pubs: *turbo folk* and Yugo-rock in Yugoslavia, *chalga* and folk music in Bulgaria, *menale* in Romania. According to him, these elements of popular culture actually 'turn the [...] picture of the Balkans upside down and convert the stigma into a joyful consumption of pleasures forbidden by European norms and taste. Contrary to the traditional dark image, this popular culture arrogantly celebrates the Balkans as they are: backward and Oriental, corporeal and semi-rural, rude, funny, but intimate' (Kiossev, 2002, pp. 182–3).

However, it also needs to be noted that the self-exoticisation apparent in films and music has not always been embraced wholeheartedly. For example, in the 1990s, when the neo-folk and turbo folk music in Serbia was increasingly associated with nationalist mobilisation and was even openly supported by the regime, one of the central traits of Belgrade rock groups was their aversion towards this type of music. Among rock groups and their fans, rock music and culture were associated with the positive values usually attached to Europe, thus to an extent perpetuating the symbolic value associated with rock music in the 1970s. Unlike some of their predecessors in the 1980s, who have experimented with mixing rock and folk music (see Ramet, 2002, pp. 135–6), these bands have rejected any association between the two music styles. As Eric D. Gordy argued, 'in contrast to its ascribed cultural value in most parts of Western Europe and America, rock and roll is perceived by Belgraders as high art and implicitly opposed to neofolk, which is regarded as "Balkan" and "primitive"' (Gordy, 1999, p. 144). Such a symbolic mapping of neofolk was shared by a large part of the urban population in Belgrade, and supported also by anti-Milošević radio stations such as B92, which refused to broadcast neofolk music. Obviously, neofolk was (and continues to be) yet another element of popular culture in the territory of former Yugoslavia which serves as a site where attachments to specific symbolic geographies are contested.

Another group of media (in a wider sense of the word) that have played an important role in spreading and contesting the new geo-cultural attachments, but have not received enough attention in existing studies, relate to food: menus in restaurants and hotels, verbal descriptions of food or references to particular dishes appearing in the various media from magazines to graffiti. Although there is no essential national cuisine, food can have a key role to play in nationalist sentiments, with invasions of 'foreign food' treated as dangerous to the national identity or celebrated for their exotic difference (Bell and Valentine, 1997). During the disintegration of Yugoslavia, a range of examples could be quoted to demonstrate such disparate uses of food. Just as elsewhere in former socialist/communist states, political changes were paralleled by a massive proliferation of new restaurants and chains serving dishes that were markedly different from those usually consumed within Yugoslavia – for example Chinese restaurants and northern American-style fast food chains. Yet the enthusiasm accompanying the introduction of these new foodscapes went hand-in-hand with the accentuation of differences between the South Slav national cuisines and the erasure of regional commonalities, in particular Balkan dishes. In Serbia, for example, the traditional *šopska* salad, made of tomatoes, cucumbers, onions and covered with grated cheese, well known in the wider Balkan region, was renamed as 'Serbian salad with cheese'. In Croatia, on the other hand, various grilled meat dishes characteristic of the Balkan region were no longer sold as Balkan, but as Croatian, and exceptions to this unwritten rule would soon be met with sharp criticism. For example, in the summer of 1996, the Croatian weekly *Nedeljnja Dalmacija* shared the shock and disgust of a Croatian minister who found out that the menu of a renowned hotel on the Croatian islands of Brioni included also a 'Balkan plate' and a 'Serbian salad' (see Rihtman Auguštin, 2000 [1997], p. 217). Following a similar logic, virtually all cafes in Zagreb in the 1990s began serving espresso and cappuccino, while Turkish coffee was increasingly hard to find in a public space (Jansen, 2001a, p. 8). In Belgrade, on the other hand, Turkish coffee would still be widely available in cafes – however, instead of being labelled as Turkish, it would be referred to as 'usual coffee' [*obicna kafa*] or cooked coffee [*kuvana kafa*].

A similar aversion to Balkan dishes could be found in Slovenia too. In a letter sent to the editor of a magazine published in Slovenia, a reader protested against the availability of Balkan dishes in Slovenia and suggested using food as a means of awakening the national consciousness: 'Let's finally get rid of the Balkan "grills" and let us make Slovenian cuisine the awakening of the conscience of those who like to eat and think in a Slovenian way' (quoted in Skrbiš, 1999, p. 120). *Burek*, a pastry containing cheese, minced meat or spinach, known throughout the Balkans, was another dish that became stigmatized in the context of Slovenian popular culture. This aversion was clearly expressed in a graffito that appeared in Ljubljana in the late 1980s: 'Burek? Nein Danke!' As Mitja Velikonja argued, 'the rejection of *burek* was meant to symbolise the rejection of Yugoslavia, emphasising Slovenia's differentiation from the other Yugoslav republics, in particular Serbia and Bosnia' (Velikonja, 2002, p. 95, n32). Still, despite ample evidence of the purification of restaurant menus and national cuisines, and the strict elimination of Balkan dishes and drinks, one should note that these trends did not necessarily affect patterns and habits of eating and drinking in the private sphere. At least some ethnographic data suggest that in their own kitchens, people continued to prepare Turkish coffee and labelled it Turkish, while private picnics are even today unimaginable without *Cevapcici* (meat rolls) and other meat dishes commonly regarded as typically Balkan. As Stef Jansen argued, this could hardly be interpreted as a conscious effort to support a positive image of the Balkans and rebel against the dominant Balkanist discourse; rather, it was simply a matter of continuity – where this was at all possible, people simply continued to live their lives in ways they were used to (Jansen, 2001a, p. 13).

Another means of communication through which the new symbolic mapping was introduced into the everyday lives of people inhabiting spaces once known as Yugoslavia were street names and monuments. Historically, few major political changes or events have passed by without being inscribed into the urban space via street-naming or monument building, and the creation of new states in the territory of former Yugoslavia is no exception. In the past two decades, several authors have examined the ideological implications of street names (Azaryahu, 1996, 1997; Pinchevski and Torgovnik, 2002). These studies have mostly remained focused on the exploration of street names as inscriptions of a particular interpretation of history, and thus of a specific collective memory, while no explicit attempt has been made to examine the street names as markers of particular symbolic geographies. The existing investigations of changes in street names during the disintegration of Yugoslavia and the formation of new independent states tend to follow the same format. In fact, most of the changes in street names marked the erasure of the earlier collective memory and the installation of a new one, and a large part of the new monuments, buildings and street names commemorated specific historical characters and events hailed by the new reading of history. The streets and squares once named after heroes, battles or brigades of the National Liberation Struggle – the struggle that served as the centre-pillar of the socialist Yugoslav unity – were now named after heroes and historical events linked exclusively to the (re-written) individual national histories (see Rihtman-Auguštin, 2000 [1997], Robinson et al., 2001). However, these renamings would, albeit in an indirect manner, mark a change in the symbolic geography as well; by commemorating particular historical events and personalities, they would simultaneously commemorate also old geopolitical and geo-cultural positionings and divisions. In some cases, the new symbolic geography was marked in a more explicit way too. For example, one of the main roads in Ljubljana, the capital of Slovenia, was renamed *Dunajska cesta* [Vienna Street], thus bringing the old capital from the period of the Habsburg Empire, and a major point of reference for Central Europe, back into the everyday landscape of Slovenians. An even more unambiguous example could be found in Zagreb, where a cinema, previously known under the name of 'the Balkans', was renamed 'Europe' (see Rihtman-Auguštin, 2000 [1997], p. 11), and where the Federal Republic of Germany and the Vatican appeared in street names, while Belgrade and Moscow disappeared (Šakaja, 2003, pp. 2). In Sarajevo, on the other hand, many new names for streets would commemorate key events and individuals from both the period of the Ottoman rule as well as the period when Bosnia was under the control of the Austro-Hungarian Empire (Robinson et al., 2001, p. 967). In such a way, two alternative symbolic geographies of Bosnia potentially got inscribed into Sarajevo streets; one positioning Bosnia as Balkan, the other as European, replicating the ambiguous positioning characteristic of public discourses in Bosnia.

Since they are subject to recurrent everyday use, street names can serve as a particularly efficient means through which a new imagining of space penetrates into people's everyday lives. Arguably, their everyday use and functionality 'mask the structures of power and legitimacy that underlie their construction and use' (Winchester et al., 2003, p. 74). Or, in Azaryahu's phrasing: 'The merit of street names is their ability to incorporate an official version of history into such spheres of human activity that seem to be entirely devoid of direct political manipulation. This transforms history into a feature of the "natural order of things" and conceals its contrived character' (Azaryahu, 1997, p. 481). Yet it needs to be noted that while changes in street names in the territory of former Yugoslavia were not, unlike the tearing down of old monuments and erecting of new ones, accompanied by special public events, they hardly passed by undisputed. The sections of letters to the editor in daily newspapers, for example, often included protests against specific changes in street names. Also, there is anecdotal evidence to suggest that new street names have not always become widely accepted, and that local people often tend to use old names long after they have been officially erased. Furthermore, as Laura Šakaja (2003) notes, the naming and renaming of streets involves official ideology, and as such it is not necessarily a good indicator of popular cultural symbolism. In order to get a better insight into the latter, she examines the imaginative mapping apparent in Croatian ergonyms – names of business associations, companies etc. – registered in the telephone directory listings in

2002. Her findings again show an overwhelming presence of references to Europe, and only a negligible number of references to the Balkans: out of 2,607 geographic ergonyms, Europe appeared in 358, while the Balkans only in three. Moreover, ergonyms including names of western European cities or countries far outnumbered those with names of eastern European cities or countries.

The list of the media that could provide valuable sources for investigations of various uses and appropriations of symbolic geographies in former Yugoslavia and successor states could certainly be extended much further, and could include cultural products such as radio and TV talk shows, TV series, stamps, banknotes, coats of arms and so on. Arguably, these media have been of paramount importance for the introduction of new symbolic geographies into everyday practices and beliefs of common people, as well as for their particular appropriations and negotiations. Although less explicit than the ones appearing in political speeches or highly profiled cultural journals, these symbolic geographies of everyday lives, to paraphrase Hillary P. M. Winchester, Lily Kong and Kevin Dunn (2003), are just as deeply implicated in the maintenance and challenging of symbolic meanings and power relations, and are worth being included into future studies.

Conclusions

Existing examinations of symbolic geographies in Yugoslavia and the new states formed in the region in the 1990s largely ignore the characteristics of sources they use to demonstrate the existence of particular mental mappings. Moreover, they tend to restrict their analysis to a narrow set of geographical labels, their uses and meanings – most often those that bear some resemblance to discourses of Orientalism and Balkanism as identified by Edward Said and Maria Todorova. As such, they do not provide enough sound bases for generalizations about the use of symbolic geographies among the mass media or political and intellectual elites in particular states, nor can they be used to make claims about the use of symbolic mappings by the whole populations of these states in their everyday lives. On the basis of issues discussed in this chapter, three main strategies can be suggested that may help overcome these weaknesses. Firstly, future investigations should consider a wider range of possible geographical labels, and pay particular attention to meanings and uses that divert from those discussed by Said and Todorova. Secondly, more thought should be given to the selection of sources, and research should include the media that are – following arguments provided above – more likely to yield examples of symbolic geographies that do not conform to the dominant ones: the Yugoslav federal media, the media with an explicitly anti-nationalist stance, various popular cultural forms and so on. Thirdly, research should focus on the uses and appropriations of particular symbolic geographies within the practices everyday life, looking at whether, how and to what extent have particular mental mappings – communicated through a wide array of the media – become accepted as taken-for granted frames of reference that organize experience on a day-to-day basis.

References

Allcock, J. B. and Young, A. (eds) (1991) *Black Lambs and Grey Falcons: Women Travellers in the Balkans*. Bradford: Bradford University Press.

Azaryahu, M. (1996) 'The Power of Commemorative Street Names'. *Environment and Planning D: Society and Space*. 14 (3), pp. 311–30.

Azaryahu, M. (1997) 'German Reunification and the Politics of Street Names: The Case of East Berlin'. *Political Geography*. 16 (6), pp. 479–93.

Bakić-Hayden, M. and Hayden, R. M. (1992) 'Orientalist Variations on the Theme 'Balkans': Symbolic Geography in Recent Yugoslav Cultural Politics'. *Slavic Review*. 51 (1), pp. 1–15.

Bakić-Hayden, M. (1995) 'Nesting Orientalisms: The Case of Former Yugoslavia'. *Slavic Review*. 54 (4), pp. 917–38.

Bell, D. and Valentine, G. (1997) *Consuming Geographies: We are What We Eat*. London and New York: Routledge.

Billig, M. (1995) *Banal Nationalism*. London, Thousand Oaks and New Delhi: Sage Publications.

Bjelić, D. and Savić, O. (eds) (2002) *Balkan as a Metaphor: Between Globalization and Fragmentation*. Cambridge, Massachussets and London, England: MIT Press.

Bracewell, W. and Drace-Francis, A. (1999) 'South-Eastern Europe: History, Concepts, Boundaries'. Balkanologie. 3(2), pp. 47-67.

Buden, B. (2000) 'Europe is a Whore', pp. 53–62 in N. Skopljanac Brunner, S. Gredelj, A. Hodžić and B. Krištofić (eds) (2) *Media and War*. Zagreb: Centre for Transition and Civil Society Research; Belgrade: Agency Argument.

Corn, G. (1991) *L'Europe et l'Orient: de la balkanisation à la libanisation : histoire d'une modernité inaccomplie*. Paris: Le Découverte.

Corn, G. (2003) *Orient-Occident, la fracture imaginaire*. Paris: Le Découverte.

Czerwiński, M. (2003) 'Discursive Construction of European Identity in Croatian Media: Recent Shifts'. *Kakanien Revisited*. http://www.kakanien.ac.at/beitr/fallstudie/MCzerwinski1.pdf. Accessed 25 March 2005.

Čolović, I. (1997) *Politika simbola: ogledi o politickoj antropologiji (The Politics of Symbols: Essays on Political Anthropology)*. Beograd: Radio B92.

Delanty, G. (1995) *Inventing Europe: Idea, Identity, Reality*. Basingstoke and London: Macmillan.

Gingrich, A. (1998) 'Frontier Myths of Orientalism: The Muslim World in Public and Popular Cultures of Central Europe', pp. 99–127 in B. Baskar, and B. Brumen (eds) (1998) *MESS: Mediterranean Ethnological Summer School*, Piran-Pirano, Slovenia 1996, Vol. 2. Ljubljana: Inštitut za multikulturne raziskave.

Goldsworthy, V. (1998) *Inventing Ruritania: The Imperialism of the Imagination*. London and New Haven: Yale University Press.

Goldsworthy, V. (1999) 'The Last Stop on the Orient Express: The Balkans and the Politics of British in(ter) vention'. *Balkanologie*. 3(2), pp. 107–116.

Gordy, E. D. (1999) *The Culture of Power in Serbia: Nationalism and the Destruction of Alternatives*. Pennsylvania: The Pennsylvania State University Press.

Guzina, D. (1999) 'Inside/Outside Imaginings of the Balkans: The Case of the Former Yugoslavia'. *Balkanistica*. 12, pp. 39–66.

Helms, E. (2004) 'East and West Kiss: Gender, Orinetalism, and Balkanism in Muslim-Majority Bosnia-Herzegovina', paper presented at the conference "Trouble with the Balkans: The First Balkan Feminist Conference, Sarajevo, 1–3 November 2004.

Iordanova, D. (2000) 'Are the Balkans Admissible? The Discourse on Europe'. *Balkanistica*. 13, pp. 1–34.

Iordanova, D. (2001) *Cinema of Flames: Balkan Film, Culture and the Media*. London: British Film Institute.

Jansen, S. (2001a) 'Svakodnevni orijentalizam: doživljaj 'Balkana'/'Evrope' u Beogradu i Zagrebu' ('Everyday Orientalism: The Experience of 'the Balkans'/'Europe' in Belgrade and Zagreb'). *Filozofija i Društvo (Philosophy and Society): Journal of the Belgrade Institute for Social Research and Philosophy*. 18. http://www.instifdt.bg.ac.yu/ifdt/izdanja/s_index. Accessed14 August 2004.

Jansen, S. (2001b) 'The Streets of Beograd: Urban Space and Protest Identities in Serbia'. *Political Geography*. 20 (1), pp. 35-55.

Jezernik, B. (2004) *Wild Europe: The Balkans in the Gaze of Western Travellers*. London: SAQI, in association with the Bosnian Institute.

Kiossev, A. (2002) 'The Dark Intimacy: Maps, Identities, Acts of Identification', pp. 165-90 in D. Bjelić and O. Savić (eds) (2002).

Kurspahić, K. (2003) *Prime Time Crime: Balkan Media in War and Peace*. Washington, D.C.: United States Institute of Peace Press.

Lindstrom, N. and Razsa, M. (2004) 'Balkan is Beautiful: Balkanism in the Political Discourse of Tuđman's Croatia'. *East European Politics and Societies*. 18 (4), pp. 628-50.

Mihelj, S. (2004a) 'Negotiating European Identity at the Periphery: Media Coverage of Bosnian Refugees and 'Illegal Migration'', pp. 165-89 in I. Bondebjerg and P. Golding (eds) (2004) *Media Cultures in a Changing Europe*. Bristol: Intellect Books.

Mihelj, S. (2004b) The Role of Mass Media in the (Re)Constitution of Nations: The (Re)Constitution of the Slovenian Nation through the Media Representations of the Plebiscite for an Independent Slovenia, Bosnian Refugees and Non-Registered Migration (1990-2001) (unpublished Ph.D. thesis, Institutum Studiorum Humanitatis - Ljubljana Graduate School of the Humanities).

Nenadović, A. ([1996] 2000) '*Politika* in the Storm of Nationalism'. In N. Popov (Ed.) (2000 [1996]) *Road to War in Serbia: Trauma and Catharsis*. Translated by Drinka Gojković. Budapest: Central European University Press, pp. 537-64.

Neumann, I. B. (1999) *Uses of the Other: 'The East' in European Identity Formation*. Manchester: Manchester University Press.

Norris, D. A. (1999) *In the Wake of the Balkan Myth: Questions of Identity and Modernity*. Basingstoke and London: Macmillan; New York: St. Martin's Press.

Patterson, P. H. (2003) 'On the Edge of Reason: The Boundaries of Balkanism in Slovenian, Austrian, and Italian Discourse'. *Slavic Review*. 62 (1), pp. 110-41.

Petković, R. (1985) Nesvrstana Jugoslavja i savremeni svet: spoljna politika Jugoslavije 1945-1985 (*Non-Allied Yugoslavia and the Modern World: Yugoslav Foreign Policy 1945-1985*). Zagreb: Školska knjiga.

Pinchevski, A., and Torgovnik, E. (2002) 'Signifying Passages: The Signs of Change in Israeli Street Names'. *Media, Culture & Society*. 24, pp. 365-88.

Ramet, P. (1985a) 'Apocalypse Culture and Social Change in Yugoslavia', pp. 3-26 in P. Ramet (Ed.) (1985) *Yugoslavia in the 1980s*. Boulder and London: Westview Special Studies on the Soviet Union and Eastern Europe.

Ramet, P. (1985b) 'The Yugoslav Press un Flux', pp. 100-127, pp. 3-26 in Ramet, Pedro (ed.) (1985) *Yugoslavia in the 1980s*. Boulder and London: Westview Special Studies on the Soviet Union and Eastern Europe.

Ramet, S. P. (1992) 'The Role of the Press in Yugoslavia'. pp. 414–41 in J. B. Allcock, J. J. Horton and M. Milivojevic (Eds.) *Yugoslavia in Transition: Choices and Constraints: Essays in Honour of Fred Singleton*. Oxford and New York: Berg.

Ramet, S. P. (2002) *Balkan Babel: The Disintegration of Yugoslavia from the Death of Tito to the Fall of Miloševic* (4th edition). Boulder, Colorado and Oxford, UK: Westview Press.

Rihtman Auguštin, D. ([1997]2000) 'Zašto i otkad se grozimo Balkana' ('Why and since When do We Detest the Balkans'). In D. Rihtman Auguštin (2000) *Ulice moga grada (The Streets of My City)*. Zemun: Biblioteka XX. vek and Belgrade: Čigoja štampa, pp. 211–36.

Robinson, G. J. (1977) *Tito's Maverick Media: The Politics of Mass Communications in Yugoslavia*. Urbana, Chicago and London: University of Illinois Press.

Robinson, G. M., Engelstoft, S. and Pobric, A. (2001) 'Remaking Sarajevo: Bosnian Nationalism after the Dayton Accord'. *Political Geography*. 20 (8), pp. 957–80.

Said, E. ([1978] 1995) *Orientalism*. London: Penguin Books.

Šakaja, L. (2001) 'Stereotipi mladih Zagrepčana o Balkanu: Prilog proučavanju imaginativne geografije' ('Zagreb's Young Residents' Stereotypes of the Balkans: A Contribution to the Investigation of Imaginative Geography'). *Revija za sociologiju (Review of Sociology)*. 32 (1-2), pp. 27–37.

Šakaja, L. (2003) 'The Landscape of Enterprise Names and Cultural Identity in Croatia', in M. A. Abreau (Ed.) Historical Dimensions of the Relationship between Space and Culture, Rio de Janeiro Conference, 10–12 June 2003 (CD-ROM). IGU Commission for the Cultural Approach in Geography.

Sirc, L. (1979) *The Yugoslav Economy Under Self-Management*. London: Macmillan.

Skrbiš, Z. (1999) *Long-distance Nationalism: Diasporas, Homelands and Identities*. Aldershot, Burlington, Singapore and Sidney: Ashgate.

Thompson, M. (1999) *Forging War: The Media in Serbia, Croatia and Bosnia-Hercegovina*. Luton: University of Luton Press.

Todorova, M. (1997) *Imagining the Balkans*. Oxford: Oxford University Press.

Velikonja, M. (2002) 'Slovenia's Yugoslav Century', pp. 84–99 in Djokić, Dejan (Ed.) (2002) *Yugoslavism: Histories of a Failed Idea, 1918–1992*. London: Hurst & Company.

Vogrinc, J. (1996) 'Close Distance: Dilemmas in the Presentation of the War in Bosnia in the Daily news Bulletin of TV Slovenia', pp. 11–18 in J. Gow, R. Paterson and A. Preston (Eds.) (1996) *Bosnia by Television*. London: British Film Institute Publishing.

Volčič, Z. (2005) 'The Notion of 'the West' in the Serbian National Imaginary'. *European Journal of Cultural Studies*. 8 (2), pp. 155–175.

Winchester, Hilary P. M.., Lily Kong and Kevin Dunn (2003) Landscapes: Ways of Imagining the World, Harlow: Prentice Hall, 2003.

Wolff, L. (1994) *Inventing Eastern Europe: The Map of Civilization on the Mind of the Enlightenment*. Stanford: Stanford University Press.

Živković, M. (2001) 'Nešto između: simbolička geografija Srbije' ('Something in-between: The Symbolic Geography of Serbia'). *Filozofija i Društvo (Philosophy and Society): Journal of the Belgrade Institute for Social Research and Philosophy*. 18. http://www.instifdt.bg.ac.yu/ifdt/izdanja/s_index. Accessed 20 August 2004.

Žižek, S. (2000) *The Fragile Absolute*, London: Verso.

Sicilian Film Productions: Between Europe and the Mediterranean Islands

Giuliana Muscio

Emanuele Crialese's *Respiro* (2002) stands at the crossroads of regional (Sicilian) cinema and the globalized media scene, offering a useful test case to explore the logics of film's relationship to cultural identity. The last sequence of the film, characterized by mesmerizing visuals, offers an underwater shot of feet moving together in a sort of ritual dance, as a mother, father and son are reunited in deep and clear Mediterranean waters during a religious celebration. This sequence generates an immediate association with *L'Atalante* (Jean Vigo, 1934), particularly the famous scene in which the husband dives in to the water to look for his spouse. The association of the two sequences is reinforced by the similarity in how the two films represent the complex relation between the couple and their sociocultural context, and in the difficult containment of sensuality. Aligning itself to the tradition of 'quality European cinema' evoked by its French predecessor, *Respiro* lacks a conventional story-line and avoids the psychologically motivated characters typical of the classic Hollywood cinema. In addition, the end of *Respiro* does not bring narrative closure to the film. Instead, rather like a dream, we are unsure whether the mother, who has disappeared, is really there, and whether the family is actually reunited in the waters in a miracle of nature and religion, or if this scene is only a projection of desires. The film is a search for a woman who has mysteriously disappeared, as is *L'avventura* (by Michelangelo Antonioni, 1960), which was also shot on a Sicilian island. Thus *Respiro* is very much in line with the tradition of quality European cinema, mobilizing cultural (and filmic) conventions antithetical to the dominant commercial cinema. But the peculiar expressive strength of the film depends on its 'Sicilianity', on its representation of the Sicilian landscape, its social conventions and culture.

In recent years the emergence of a 'Sicilian cinema', marked by an outstanding number and quality of films, coincided with a larger trend towards the regionalization of film production in Italy.[1] Strongly 'European' and 'Sicilian', *Respiro* is less prominently Italian in its cultural profile. While often associated with the neo-realist tradition, its choice of mythological (and anthropological) rather than socio-historical dimensions weakens its connection with this Italian style. For instance, the film deals with Sicily without any reference to the mafia or to the current socio-economic problems of the island and, in so doing, sidesteps Italian stereotypes about Sicilian lifestyles. *Respiro* is directly connected to the tradition of an anthropological approach to Southern Italian culture, with its attention to superstition, rituals, gender roles, and specifically, to the Mediterranean Mother. To further complicate the argument of the weak 'Italian' identity of *Respiro*, the director of the film, Emanuele Crialese, studied cinema in New York and claims greater familiarity with American cinema than with Italian film culture (Bertani, 2003). While most non-Italian critics do their best to find similarities between protagonist Valeria Golino (and her character), and Anna Magnani or Sophia Loren, the director places himself outside, if not against, the Italian film tradition, and particularly its neo-realist associations.

This ambiguity in the national identification of a media product is not at all rare today, particularly in the context of globalized production and communications. However, *Respiro* represents an interesting case study because of its international visibility, and because it poses significant cultural questions.

We will try to examine the issue of its identity on both the cultural and the production levels, remembering of course that they are inextricably connected and continuously repositioned by the point of view of the investigation. In fact the identification of a film as European, national and/or regional depends on an interplay of institutional practices, both critical and operational, since there are multiple agencies legitimizing these definitions, representing conventional critical categories (as in the case of national cinemas) but also technical production contexts, as in the case of the access to European funds in a transnational co-production.[2]

The 'Sicilian' film *Respiro,* with its cultural and production logics simultaneously bound up in regional, Italian and European contexts, forces a re-consideration of the geographic identification of media products, encouraging a re-examination of definitions and assumptions in this respect.

Global or international?

The concentration and internationalization of the media economy tends to create culturally standardized products – globalized products – which travel well on the television screens of the planet, satisfying a mass audience's demand for audio-visual entertainment, but not necessarily addressing the intellectual curiosity of educated and cinephilic audiences. This creates a *niche* audience for a less standardized cinema, that is, for 'non-Hollywood' or 'non-Hollywoodized' films, which present uncommon scenarios, overlooked cultural traditions, and important social themes. This cultural need is confirmed by the international success of Korean melodramas and other Asian films, rich in sex and violence or proposing a choreography of martial arts, spectacular in a way which is quite different from Hollywood standards, as well as by the popularity of Iranian fables or of shocking social reportage.

As international as the globalized audience, this audience for so-called quality cinema is catered to by film festivals and specialized theatres. It sustains an international film market, of smaller proportions in terms of revenue than the commercial circuit, but of evident cultural impact for the critical discourse it stimulates. This international film market grew out of the alternative distribution channels of university film clubs and art-houses in the US as well as in Europe, but it is no longer focused mainly on European and auteur cinema. Its international scope has expanded, and its interest in authorship has decreased in favour of a broader curiosity about cultural diversity. While it is evident that the consumption of media products is less and less defined and contained by national borders, we should distinguish between a globalized audience, consuming standardized commercial products of various quality produced by Hollywood or by multinational companies, and an international – perhaps even cosmopolitan – public, more selective in its taste, and more attentive to transnational niche productions. The juxtaposition global/regional (and commercial/innovative) seems to have emerged in place of the traditional media dialectic between American and European cinema, which represented the dominant critical and distribution trends throughout the Sixties and Seventies. Globalization seems to have encouraged local reactions, which re-positioned cultural productions in a new set of geo-cultural relations, creating a dialectic between national (and international) cinema and regional films.

This study will focus on this paradigm shift, re-examining the crisis in traditional geo-cultural relations. Within the global media system, we can try to narrow the focus in geographic terms: the problematic definition of European cinema first, a socio-historical view of Italian cinema of the last decade next, and in the final section, Sicilian film, as regional cinema – a 'localized experience' – *Respiro* as a case study.

European film – the historical definition

Which cultural, commercial and technical criteria define a European film? The historical and critical debate usually juxtaposes stereotypes of the American and European cinemas, connecting their respective cultural impact and commercial strategies to the socio-economic context and to the scene of international affairs.

In *Mass Culture and Sovereignity,* Victoria De Grazia analyses the first wave of Americanization following World War One and defines the differences between the two cinemas:

> The American cinema stood for major economies of scale, capital-intensive technologies, and standardization; it favored an action-filled cinematographic narrative focused on the star and pitched to a cross-class audience...the European tradition was identified with decentralized artisan-atelier shops and was associated with theatrical and dramatic convention attuned to well-defined publics. It rested on a commercial network mediated

> by intellectuals – meaning directors, technicians, and actors...as well as cultural organizers and critics. (DeGrazia, 1989, p.61)

Andrew Higson and Richard Maltby in their introduction to *Film Europe and Film America* add an important reformulation to this definition:

> The idea of a single, coherent European cinema does not do justice to the facts. Viewed in terms of reception as well as production, European film culture is perhaps best understood as a series of distinctive but overlapping strands. One strand is the Americanised metropolitan popular culture, confidently modernist, knowingly part of the emergent consumer culture and designed to travel. Another is the parochial or provincial non-metropolitan culture which embraces local or national, rather than international popular traditions, and which is often inexportable...A third strand to European film culture draws on the high culture of 'old Europe', and embraces much of what we now understand as art cinema and heritage cinema; again, this is metropolitan based and designed to travel.' (Higson & Maltby, 1999, p. 20)

The definition of 'European film', consistently recognized as problematic, operates on two levels: one critical and cultural, mainly presented in contrast to Hollywood; the other economic and institutional, often related to the defensive measures against the dominance of American audio-visual products in Europe, or to the characters of film production in Europe (state-assisted, quality-oriented, and encouraging transnational co-productions).

After the epochal changes of the 1990s, in face of the formation of a larger Europe, De Grazia complicates the picture:

> If the perspective is global, might not the European cinema be linked to some notion of western as opposed to eastern or colonial cinema; should it be juxtaposed to African, Latin American, Indian or Asian cinema? Finally, if there is a European cinema, what exactly is its relationship to the 'national' cinemas within Europe; what qualities link together, say, art productions and outright commercial products, films of evasion and propaganda works, to group them with their equivalents in adjacent societies to form a distinctive European amalgam? (De Grazia, 1998, p. 7)[3]

Therefore, in fewer than ten years, De Grazia's clear-cut juxtaposition between American and European cinema has been replaced by a global media sphere, challenging the relations between national and European cinemas, but also between 'colonial' and 'imperial' film cultures, and the very concept of national paradigms.

After the economic and political unification of Europe, there were great expectations for fundamental changes in film culture and a stimulation of European film production.[4] However, while Europe is a very concrete transnational institution, the identification of a cultural product as 'European' is rare, with national identification still prominent. But cinema has been granted a crucial role within the debates on European identity, and this centrality further motivates an investigation of the characteristics a film must have to be labelled 'European'.

In her analysis, De Grazia (1998, pp.27–9) insists on the crucial role of the European media sphere: 'European governments tacitly concurred on dividing the audiovisual sector according to two models: the movie industry following a market model while television remained a state monopoly.' She notes that, 'this division deterred them from conceiving of any over all national or transnational audiovisual politics', assuming 'that a modernised economy would yield a common national and cross-European value system'. In her opinion, this European project failed – a commercial defeat that is particularly negative in her view, given that 'by the mid-1960s, the then nine members of the European Community comprised an audience and a productive capacity larger than that existing in the US'. De Grazia observes how 'the linkages provided by a European-based audiovisual industry would appear as central, with communication, culture and information reinforcing identity across the common territory', lamenting instead 'that not even a unified Europe can easily control this audiovisual space because it is dominated from without.' Globalization has provoked a crisis of the state-supported, public service, and quality-oriented media production of the European tradition, thus weakening the traditional identification of its 'Europeanness'. This is particularly evident in the histories of the national broadcasting systems within Europe, which have seen their national identity seriously undermined by growing commercialization.

While there is no doubt that globalized media have conquered commercial theatres and television programming in the old continent, it is also evident that there are old and new signs of 'resistance' both in film production and in exhibition, for instance with the operation of the Europa Cinemas circuit.[5] The international success of quality (and yet popular) films made in Europe, as in the case of *Le fabuleux destin d'Amélie Poulain* (Jeunet, 2001) or *La vita è bella* (Begnini, 1997), and *Respiro*, keeps the concept of European film alive, even though the label 'European' is rarely used to refer to them in journalism. Our case study confirms this tendency, in that an investigation of the terms used in American popular daily papers reveals that the reviewers tend to label *Respiro* 'Italian', and never use the term 'European'.[6] The category most commonly applied to 'non-American' cinema nowadays is 'foreign films', or 'international cinema', which normally includes the European quality product.

The economic, political and cultural relevance of the media question in Europe is obvious. In the slow transition from national policies to a unified European media strategy, and taking into account the infrequency of the term 'European' as a commercial label, we should start reconsidering the relation between European, national and regional film production.

Apparently the (recurrent) experiment in the creation of a European film cartel, be it the Twenties' project of German inspiration, or the rich experience of transnational film co-productions of the Sixties, or the French resistance in the GATT (General Agreement on Tariffs and Trade) negotiations, is less prominent in public discourse today. On the other hand there is a concrete 'Europeanization' of media productions through an array of policies and institutions. There are several offices and agencies in Brussels and Strasbourg working on media policies and supporting film production, such as Eurimages and the European Media Commission. Interestingly enough, there is also a plurality of experiences in which the European Union supports local media policies.

The activities of these agencies have markedly influenced film production in Europe, but their impact does not seem to have been registered by the public. The logos of these European institutions, which regularly appear at the beginning of film credits, seem to contribute only mechanically, formally, to the creation of the film. These labels in blue, with their golden stars, lack the mythical resonance of the well-established logos of the American majors with their legendary connotations; their cultural significance in terms of Europeanness is still weak. And yet they signify an important production factor: the film we are about to see was made with European support – an element which actually affects its cultural identity in ways that go beyond what the audience can see.

From the national context to regional production

In these days of media globalization, European television is only institutionally 'national', while it speaks a global language. Europe's national film industries, on the contrary, have moved both in the direction of localization, with the institution of regional film commissions and an escape to regional locations, and in the direction of European integration, with the creation of the European Media Commission, with its important role in supporting transnational co-productions and the film economy in a more general sense. The 'European tradition' of television, identified by the motto 'to educate, to inform and to entertain' has been reversed, with entertainment coming first. The linkage of American television with the commercial model, and of European television with public service, is not that clear-cut anymore. Nowadays Italian RAI functions as a public service in the 'European tradition' only in off-hours, while most prime time programming consists of foreign formats and imported television series – globalized as much as the programming of the commercial networks. A possible exception is represented by quality made-for-TV fiction, which is culturally 'national', both in themes and narrative contexts, often adapted from national literature, and shot on provincial sets (Naples for *La squadra* and *Un posto al sole*, Middle Italy for *Maresciallo Rocca*, Sicily for the Montalbano series). In addition to representing the prestige product of the networks, this markedly 'Italian' fiction also aims at the international market, thus exploiting a variety of Italian landscapes and regional cultures, in contrast to the standardized studio sets of most televised fare.

Contemporary Italian film production seems to move instead in a different direction – closer to the state-controlled model and to the philosophy of quality-production and public service traditionally associated with European television. Since the Seventies, a system of tax deductions and state funds has supported the production of quality films in Italy through different institutional strategies. This policy has allowed the emergence of a new generation of film-makers, less 'authorial' than in the past, but narratively innovative and geographically diversified.[7]

In the light of the tripartition offered by Higson and Maltby (1999) in reference to European film, we can identify samples of all three categories within contemporary Italian cinema. The presence of 'the Americanised metropolitan popular culture, confidently modernist', which normally includes international stars, as in the case of *Non ti muovere* (by Sergio Castellitto, 2003) with Penelope Cruz, is still present. The production of 'art cinema and heritage cinema', again 'metropolitan-based', remains a major part of the national film output, with a platoon of 'authors', from Bertolucci to Amelio or Moretti. Less evident is the 'parochial or provincial non-metropolitan culture which embraces local or national, rather that international popular traditions, and which is often inexportable' (Higson & Maltby, 1999). This type of product now belongs to television, while regional films with incomprehensible dialects and the representation of local cultures have merged with art cinema, circulating in festivals and art houses. Films such as *LaCapaGira* (by Alessandro Piva, 1999) or *Sangue vivo* (by Edoardo Winspeare, 2000), presenting musical and anthropological traditions of the South such as *la pizzica* (the traditional dance which developed into the *tarantella*), fall within the tradition of *La terra trema* by Luchino Visconti or within Pasolini's cinema in their use of non-professional actors and local dialects. These films are both regional and international, but never 'provincial'. The first element to be taken into account in our study, then, is that nowadays, regional film is not necessarily a provincial or conservative cultural product.[8]

Contemporary Italian cinema includes a few international products, diversified enough in their production values so as not to be 'globalized' and yet capable of a good commercial performance. Examples include *La vita è bella*, *La stanza del figlio* (Nanni Moretti, 2001), and independent regional and 'authorial' productions such as Mario Martone's Neapolitan cinema. Today's big made-in-Italy money-makers are the big-budget 'Christmas comedies' produced for mass consumption and catering to an audience accustomed to the facile rhythms of television. Globalized in their standardized formulas, they apparently represent national culture, but are too provincial for international audiences while being totally unrelated to the traditional qualities of 'Italian cinema'. In a way, this 'globalized' product now represents the provincial film taste earlier associated with regional cinema. The axes international/global and regional/provincial alter and complicate the classification of contemporary Italian cinema in interesting ways.

These changes in the media market are the product of both economic and sociocultural conditions. In the last decade, Italian film and television production have been concentrated in the hands of a very small group of companies, evident in the dominant position of RAI and Mediaset in television, and that of Medusa Film as a producer-distributor and exhibitor in cinema, all owned by media tycoon and Prime Minister, Silvio Berlusconi. The output of this system is usually a globalized product conforming to 'commercial standards', appealing to large audiences, and targeted to both big and small screens.

At the same time, socio-political conditions encouraged localized culture. On the political level there are regional (and separatist, intolerant) movements such as the Lega; in a more general and positive sense, there is a movement for preserving regional differences in language, traditions, and food, which has received the support of the European Union.[9] An analogous phenomenon of regionalization is happening in film-making. The institution of regional Film Commissions encouraged film-makers to make their films in different regions, with the support of these local agencies. In addition, regional Film Commissions often use European funds to support educational activities, such as local film schools or local initiatives in media education.[10]

Regionalization is not limited to the institutional sphere of media production. Whereas mainstream Italian cinema and television fiction are shot in Roman studios, since the 1980s the production of more culturally ambitious films has increasingly been made in other regions and *en plein air*, in real spaces. In the neo-realist tradition, film-makers leave Cinecittà, rediscovering the Italian regional and provincial landscapes. This movement signals a desire for more independence, and for a closer contact with regional experiences. Thus, next to the more conventional Rome-centred media production, there is a proliferation of local experiences, including the 'Milanese', the 'Neapolitan', the 'Tuscan', and most of all, the 'Sicilian' cinema.

Sicilian cinema represents a chapter of Italian film history, and according to its filmography, a very relevant one.[11] It is characterized by an emphasis on landscape, on insular anthropology, and is often devoted to specific themes and issues (mafia, hard labour in fishing, in the fields, in the mines). Traces of old and new social problems filter through even the more personal of these films: strong family ties, the submission of women to the patriarchal order, the difficult socio-economic conditions, and the mafia.

This cinema makes implicit references to the 'Southern Question', the socio-economic issues that characterize the relationship between the industrial North and rural South in national history. Since the unification of the country, the North has always held a dominant institutional position, while Sicily has distanced itself from the 'continent', its insularity periodically verging on separatism.

In recent years there has been a large output of Sicilian films. The list includes *Cinema Paradiso* (1988) and *Malena* (2000) by Sicilian film-maker Giuseppe Tornatore, national productions such as *Cento passi* (2000) by Marco Tullio Giordana, *Placido Rizzotto* (2000) by Sicilian director Pasquale Scimeca, *My Name is Tanino* (2002) by (Sicilian-born and Tuscan-educated) Paolo Virzì, *Tano da morire* (1997), *Sud Side Stori* (2000), and *Angela* (2002), by Milanese Roberta Torre (who lives in Sicily), the films by Sicilians Ciprì and Maresco, *Perduto amor* (2003) by Sicilian composer Giacomo Battiato, *L'isola* (2003) by Sicilian Costanza Quatriglio, and *Respiro* by Milanese-American Crialese.

These films reformulate the geography of the national imaginary, re-organizing the space of the narrative: characters coming from the North move southward, to Sicily – almost to Africa. But their movement is also a movement in time, connecting with History and Myth. Artistic and literary references, monuments, architecture, or direct narrative elements emphasize the deep and articulated roots of the insular culture.

In the cultural history of the island, the soul of Sicily is divided in two parts: the Magna Grecia, that is, the roots of ancient western civilization, in the East, and the Arab (but also Norman) part of the island, with Palermo as its capital, in the West. Thus cultural elements referring back to Italy, even to the ancient Roman Empire, are quite weak in Sicilian heritage. Sicilian cultural roots connect eastwards with Greece and with the ancient heart of the Mediterranean, or southwards, with Africa. The North represents the destination for migrating labourers, or the force of exploitation, but also an intermittent aspiration to modernization and liberation from an ancient self-imposed cultural conditioning.

Nowadays Sicily is 'invaded' by 'boat peoples' coming from the very cultural areas of its mythological origins: from the Ionian Sea and from the coasts of North Africa, which explains the tensions associated to its insularity. Sicily is perceived as a natural bridge from Africa and the Middle East to Europe by desperate populations in search of re-location who pass through the island, not to inhabit it, but to move North. Perceived as a piece of Europe by North Africans and Easterners, Sicily is considered 'African' by the 'continentals'.

As seen in the news and in the papers everyday, the geopolitical position of Sicily is the source of a continuous crisis, which creates a traumatic conflict between its mythological past and the socio-economic problems of peoples who are actually deeply connected with those cultural roots, now looking for a future. Travel to Sicily can today be a beginning or an end, a return or a passage, the past and the future, because of the sociocultural implications of its geographical position. Therefore, in addition to belonging to a historical tradition, Sicilian cinema might have more specific contemporary determinants due to the symbolic space the island can occupy in the articulation of an imaginary North/South, but also East/West, and its peculiar a-Italian culture.

As mentioned, there are also more media-determined explanations for the growing number of regional films. The process of globalization and the reaction of regionalism and cultural fragmentation encourage the adoption of circumscribed identities. Because of their defensive stance, regionalism and localism are often associated with conservative instances and provincialism, but we have already argued that localism is not necessarily synonymous with these backward trends in contemporary media production. Given that insularity and separation imply the need to fill the distance – the gap – in communications, at times this need stimulates innovation.[12] Sicilian cinema confirms this innovative potential of regional production, both in its working methods and in themes. These films are quite experimental at times, within an interesting dialectic between cultural traditions and innovative style, as in the work of Ciprì and Maresco, Roberta Torre, Pasquale Scimeca, and Crialese.

As we suggested, in their narrative, the direction of the travel is frequently inverted: while in neo-realist cinema, characters travelled from Sicily to the North (as in *Paisà* by Roberto Rossellini, 1946, and *Cammino della speranza* by Pietro Germi, 1953), paralleling the coeval migration of Southern Italians to the North in search of a job, since *Ladro di bambini* (Gianni Amelio, 1992) Sicily has become the place of a return, not a place to be abandoned. In the 1980s, many film characters travel from northern Italy (perceived as an industrialized, alienated, and globalized space) in order to explore the South as the site of a more authentic identity – in order to return to a sort of 'state of nature' as in the last, beautiful, images of *Respiro*. In *Ladro di bambini* and *Respiro*, the travel to Sicily ends with somebody diving into the sea, in a liberating *lavacre*. It is a dive in search of a Nature that in Sicily seems still in touch with Culture, through Myth.

Destination – escape – return: Sicily allows for all the directions of travel.

A Sicilian film: *Respiro*

Within the domain of Sicilian cinema, *Respiro* is a perfect case study, both for its thematic elements and its mode of production. It tells the story of Grazia, a woman so close to nature that not even her family accepts her vitalistic spontaneity. She is animal-like and almost incestuously close to her young boys, two *scugnizzi* who love to fight with the other kids. She also has a precocious adolescent daughter. Her husband is in love with her but does not understand her and lets family and friends convince him that she must be 'sick'. Considering her mad, they plan to have her 'cured' in Milan. The woman rebels and with the assistance of her eldest son, runs away. She hides in a cave on the rocks overlooking the sea, but she is believed dead, provoking the community's grief (and sense of guilt). Even her son cannot find her anymore, but in the final sequence, during a religious ritual in the village, her husband and sons dive in the sea and join her in the watery deep.

From a thematic point of view, we should link general notions about the insularity of Sicilian cinema with the fact that the film was shot in Lampedusa – an island next to the island of Sicily. Thus *Respiro* doubles its 'insularity', and the possibility of creating distances – removing the narrative from geographic (and historic) parameters. Although shot in Lampedusa, the very island which is today the main stop of clandestine immigration, the film makes no reference to this contemporary crisis in its narrative. The film instead reworks its complex geographical implications, playing on the time element, moving into the a-temporality of myths and fables, escaping contemporary social reality by diving with nymph Grazia into the deep clear waters.

The main characters of *Respiro* are not adults or responsible people, but adolescents or 'immature' personalities such as Grazia (which means Grace). Reality is seen through their eyes – directly, with a naiveté that avoids preconceptions and becomes magic realism, allowing for a direct approach, the depiction of landscapes and the anthropological description of moments of collective life, without creating a realistic rendition. There is a suspension of time and verisimilitude, giving the film a dream-like quality. And yet, most reviews seem to notice instead the film's documentary-like quality, without questioning the incoherent coexistence of realistic and surrealistic claims in the text.

According to some interviews, the director deliberately constructed the timelessness quality of the film in order to keep the audience guessing about when the story was actually set. Grazia listens to Patty Pravo on a portable 45rpm record player and young people walk for the *struscio* on Saturday, the evening stroll typical of Rome – behaviours which seem to set the story back in the Sixties. The decor and the costumes favour a sense of timelessness, but this strategy also connects with the expectation that in isolated Lampedusa, behaviours and clothing styles could be outdated. As Crialese said: 'In the end this is what myth really is: when you take away all the details; an analysis of human events and human feelings which are eternal' (Bertani, 2003, p. 61).

From a thematic point of view, the tension between the Mediterranean matrix of the woman and the repressive forces of medicine and the police associated with men, masculinity, and with the North, emphasizes a value conflict, but also a symbolic contrast. The mother finds the solidarity of her son, and, in an indirect way, of her daughter, who seduces a rigid northern policeman with her natural grace, taking him away from his duties. When Grazia disappears she becomes a legendary figure, not the victim of an accident. In fact, in the second part of the film there is no clear distinction between 'reality' and the suspended state, embodied by Grazia and her son, and their special view of the situation, and of the landscape.

The cave and the bay where she takes refuge represent a separate space, a sort of island within the island. Grazia's space is the domain of a new mythology, not a geographic space. It echoes ancient myths: Circe, the nymphs or the sirens, who attracted men, distracting them from duty and adventure, plunging them into a marvellous and yet traumatic immersion. In the film, men are seen as primitive warriors, connected to fishing – an ancient profession, but also hunting the wild dogs that Grazia has liberated in the streets, and shooting them from the roofs of the village in a barbaric eruption of violence which resembles a war. The children are organized into gangs and fight as little wrestlers, in an embrace that mimes both the fight and the sensuality of their young bodies; in an expression of adolescent aggression and tenderness. In fact *Respiro* is rich in situations that are narratively and figuratively ambiguous, polysemic, making the viewer wonder about what is actually represented on the screen, as when Grazia lies in bed with her younger son.

The image of the sea is crucial in this film, because it is not seen as a space of infinity, as a shimmering, endless horizon, but always in its interaction with the island, often from above or from beneath: a body of water into which one can dive. All these associations implicitly suggest suicide by drowning, remaining without breath, whereas the very title of the film is breath, *respiro*. But diving also implies a positive contact with a primal element – water – the idea of purity, of ablution or bathing, and the desire of swimming freely as an ancient aquatic creature.

Interestingly enough, other Sicilian films are inhabited by nymphs, ready to dive into the water when a sacrilegious man is unwilling to recognize the secret ties between the island, the sea and the sun. In *My Name is Tanino* the American girl dives and swims in the ocean with the grace of an irresistible siren, attracting the boy in her dangerous (social) environment. When these women dive into the water, the story regresses to the time of Myth, outside History. And yet they signal precise cultural formations – Nature and Culture, South and North, Africa and Europe.

The anthropological culture at play throughout the film is connected to the traditions of Sicilian cinema, but also to Crialese's stated intent of disassociating the film from Sicilian folklore and stereotypes, and reverting to a less conventional and more authentic representation of the island. The conflict between Culture and Nature, contemporary civilization and social rules against sexuality and a state of nature that Grazia physically embodies, address issues of both universal value and ancient resonance.

From an expressive point of view, *Respiro*'s apparent visual directness (and beauty) is actually an eclectic melange of different film styles and traditions. It combines the documentary approach of neo-realism (for example, in the sequence in the 'factory', when women are curing the fish), with the representation of landscape as a character, more typical of Antonioni's work or authorial cinema. And yet this significant presence of the landscape differs from the neo-realist tradition, in which landscape was represented as a social space, or at least marked by the presence of men. In *Respiro* instead it is water and rocks – a symbolic representation of a primal nature, similar to the a-historical use of landscape that Pasolini evoked in *Medea*, with its mythological implications of Greek ascendance. *Respiro*'s fables and symbolic subtexts introduce a different perception of reality, a magical realism that reminds us of the Rossellini of *San Francesco,* or of the animistic representations of African cinema. In fact the use of Mediterranean colours, radiant and vivacious, the blues and yellows, the sun-drenched landscapes of *Respiro*, do evoke African cinema.

Filled with heterogeneous filmic echoes, *Respiro* actually retraces the ancient roots of Sicilian culture – the Greek mythology and African animistic presentation of Nature, becoming multiethnic in its style. But at the same time, it does not directly represent a multicultural social situation or the social problems of contemporary Lampedusa.

Sicilian cinema often moves back to ancient roots, to ancient religious rituals, like the processions that obsessively run through these films, and the primitive religious celebrations organized in the final part of *Respiro*. However the film's representation of these celebrations is not in line with the anthropological explorations of the island's religiosity in the tradition of Fifties' documentaries. On the contrary, it is its primitive, ancestral quality that seems to attract Crialese – a film-maker with a composite background.

Emanuele Crialese studied cinema at New York University, thus absorbing a non-Italian idea of film-making and a diverse film culture. His first film, *We Were Strangers* (1997), made in the US, deals with a clandestine Sicilian worker in North America and was presented at the Sundance Film Festival, a main venue for independent cinema. American-educated but not 'Hollywoodian', Crialese wanted to make a film in Italy, but not in Rome – not in Cinecittà. He states that he wanted it to be about an island – a space with a suspended time, surrounded by nature and sea, which 'happened' to be Lampedusa.[13]

In Crialese's view, *Respiro* is the story of how a community reacts to diversity, of how any community needs to find an external enemy. Therefore Grazia is the 'other'. But if we argue that she embodies Nature, the Mediterranean Mother, the deepest roots of Sicilian culture – the eastern and African ones – it is possible to make her image overlap with the untold (and unseen) enemy: the clandestine immigrant, coming from Africa and the Middle East. And even though this interpretation would take Crialese's words too far, it is textually inscribed in the film, in its multicultural style.

Notes about the production of *Respiro*

Respiro has gained notable international visibility, and has won several prizes at festivals, including the prestigious Semaine de la Critique in Cannes (which had not been won by an Italian film for forty years). Nominated for the European Film Award in 2002, it has also attracted wide critical attention. These commercial and critical successes are even more impressive if we consider that *Respiro* is a small independent film by a young director whose name is almost totally unknown in Italy and abroad. One could argue that it has achieved this success because of its artistic merits, but we know all too well that artistic qualities and thematic relevance are no guarantee of success on the film market, given the complexities of international film distribution.

What distinguishes *Respiro* within the 'Sicilian group' is the fact that it was a French-Italian collaboration, produced by Domenico Procacci, the brilliant head of independent film company Fandango, with Les Films des Tournelles, Roissy, Rouse, TPS Cinéma, Medusa, and Telepiù, and with the support of Eurimages. Favoured by its status as a European production, *Respiro* enjoyed international distribution in France, Germany, Belgium, and the US. The other Sicilian films are national productions, made both by big companies such as Medusa, or by smaller concerns, but without European funds.

This element of differentiation requires some attention, taking us back to the question of definitions. Eurimages was instituted in 1989 by the Council of Europe 'to encourage European co-productions of films and television fiction'. The intent of the European legislators was to 'support works which reflect the multiple aspects of European society and reveal how they belong to the same culture'. However, our analysis of the themes and narrative structure of the film has not traced evident signs of its 'Europeanness', except in a broad cultural sense. Instead, other goals of Eurimages seem to have a closer relation to the actual case of *Respiro*. Eurimages is intended to 'financially support and protect the European film industry from transatlantic competition', and, more specifically, to assist the production of a film, when 'at least two independent producers of at least two members states of EU participate' in its realization, which was indeed the case of *Respiro*.

An economic criterion is emphasized in the definition of 'European films', as stated in the official booklet on media financing printed by European authorities in 2003. European films are 'works produced in the majority part by enterprises with a basis in one or more countries participating in the MEDIA Program'. Or they could be 'works whose realisation significantly contributed professionals who are citizens/residents of the countries participating in the MEDIA Program'. These mechanical definitions of European film attest to the failure of establishing a common understanding of 'European identity', by reverting to the 'national identity' of the enterprises or individuals involved in the production in reference to the EU.[14] The association of the definition with transatlantic competition and with its protectionist instincts is also a sign of the difficulty of defining a European media product, if not in contradistinction from American cinema. On the other hand, the same clause includes a reference to the support of media production on European land – which is precisely what has helped the production of *Respiro*.

By supporting independent producers, Eurimages helps to create small films, which find new commercial possibilities because they are transnational co-productions. An initial flop in Italy, *Respiro* was distributed in France by its French producer, where it found critical acclaim, allowing it to return to Italy and receive wider international distribution.

Thus it is evident that a film production supported by European funds has automatic and relevant advantages on the film market because it allows a film to receive a wider international distribution. And yet the involvement of Eurimages in the making of *Respiro* appears more like a technicality than a cultural definition. The main condition required to receive Eurimages funds is that a film must be produced by companies based in at least two European countries, but there are not special narrative or thematic rules which make it 'European' as a text.[15] European in its financing and production, *Respiro* is not directly European in its themes or in its reception. This can be established by analysing the reception of *Respiro* on the American market, where previously 'Europeanness' was a possible marker of product identification. To quickly test the definition, I visited an Internet site which offers a synthesis of American popular film reviews published on American daily papers.[16] The site includes the film in the category of 'Foreign film'; and none of the reviews uses the word 'European'. The term 'Italian' often appears in the titles of these reviews; specific references to its geographical setting (Sicily, and the island of Lampedusa) are frequent. Therefore, whereas in all probability in the Seventies such an unconventional film would have been easily identified as 'European,' this term is never used by any of the 75 reviews appearing on the website. It is interesting to note that the criticism of the film apparent in these reviews refers to the 'weakness' of the plot – a misrecognition of the director's narrative strategy. Given that a traditional characteristic of European cinema is precisely its 'otherness' in relation to American cinema, particularly regarding its innovative forms of storytelling, this misreading of *Respiro*'s peculiar narrative structure amounts to criticizing it for not being Hollywood-like.

The American reviews often associate the film with neo-realism ('old fashioned Italian neo-realist simplicity'); Valeria Golino is even compared to a 'streamlined version of Anna Magnani in her prime'. In the international press any Italian film tends to be criticized according to either authorial paradigms or neo-realist traditions, even if, as in this case, the young director Crialese would refute this forced attribution of neo-realist traits, which are indeed absent in the style or in the narrative of the film.

While the America popular press does not seem to identify *Respiro* as a European film, it does define it as Italian, and strongly associates it with the conventional idea of 'Italian cinema'. However the very definition of 'Italian film' appears to be quite vague – neither historical nor geographical. For instance, one comment notes: 'it reflects the effortless charm of a film like *Il postino* rather than the untidy manufactured romance of another *Captain Corelli's Mandolin*'. The two films cited here have a different (and loose) association with Italy. *Il postino* was shot in Procida and in the Sicilian island of Salina, with an Italian actor (Massimo Troisi) and a French star (Philippe Noiret) by a British film-maker (Michael Radford) within a French-Italian co-production. *Corelli* is a multinational film production (American-French-British), with Greek locations and a character who is Italian. And yet a film shot in Mediterranean locations with an Italian 'mood' and certain thematic elements, is automatically considered to be Italian by some film critics. In our geographic game, this broad application of the notion of 'Italianicity' for a text and the consequent weakness of the definitional power of 'Italian film'' should not be underestimated. The film titles associated with *Respiro* also indicate another significant element of reception: Italian films seem to enjoy greater international popularity, especially on the American market, when they 'look' like the old Italian cinema people enjoyed in the Fifties and Sixties. Films such as *Il postino* and *Cinema Paradiso*, set in the Fifties, and *Respiro*, which seems to be set in the Sixties, have traits of a-historicity which allow them to be associated with the neo-realist tradition, in themes and visual style. On the contrary, however, *Respiro* is not a nostalgic recollection of the neo-realist tradition, but a modernist media project.

While this misconception refers to a vast gap between the views of critics and film-makers, what is relevant to our discussion is the knee-jerk cultural stereotyping that goes on when we use such terms as 'American', 'European', 'Italian' and 'Sicilian'. Our case study and definitional efforts instead demonstrate that we must be aware of the complex (and evolving) relations among these terms, and most of all of their new implications in the global market. A 'European film' can be a number of things, but in this case it is evident that the only 'Europeanness' that really helped *Respiro* was that of its production circumstances, through Eurimages. The European assistance to media production in the old continent is actually producing a major impact on the media scene. And if a small quality film by an unknown director can reach most European and American screens, than that impact can only be described as cultural.

Internationally perceived as Italian, *Respiro*'s mode of production is marked by its Europeanness, while its themes and expressive qualities are strongly Sicilian. In a way, *Respiro* is a 'cosmopolitan' artefact, simultaneously embodying both European and regional identities.

The paradigm shift Ulrich Beck identified with the concept of cosmopolitanism finds a possible application here (Beck, 2004). Refusing globalization and standardizing practices both in cultural and media terms, the film proposes a cinematic language that antagonizes the Hollywood narrative system with its open ending, echoing different the traditions of magical realism and surrealism.

Post-national in its vague Italian identity, *Respiro* takes into account the hybrid origins of the culture it represents, accepting its Mediterranean complexity and Sicilian multiculturalism. The boys struggling playfully but aggressively at the beginning of the film are statuesque and sexually ambiguous like young Greek warriors, but also metropolitan members of a street gang when they fight in the post-industrial landscape of an empty cement pool. The integration of the blond northern Italian policeman with the Sicilian girl reconciles cultural polarities of ancient ascendancy. It offers a cosmopolitan representation of Sicilian culture, which moves in time and space, allowing for a mobile identification of cultures and relations. Transnational and cosmopolitan dimensions do not substitute, but instead redefine and complement national and local identities. European cosmopolitanism could be an answer to globalization, with its positive refusal of roots, borders and fixed identities. The Mediterranean imagined space is multiethnic by definition. Its borders are impossible to define, as the undeclared wars of fishermen with national authorities easily demonstrate. There is an ancient history of multilayered identities in Sicily, which continuously surfaces in *Respiro*'s style. The final sequence is presented from an impossible point of view: from inside the sea, from the bottom of the Mediterranean. It is the vision of an (ancient) aquatic creature, un-breathing in the deep ocean, aspiring to a new fusion of the Woman, the Man and in the Son, in the very element that brought life to the planet – water. Grazia's dream in the film is a typical cosmopolitan aspiration: to live her own life, while living together with her community, in harmony with Nature.

Notes

1. The article notes: 'Sicily', as writer Vincenzo Consolo said, is 'too damned photogenic,' a quality that has often prevented the directors who approached it from communicating its truth. With a few significant exceptions, the views of these film-makers have often been superficial or stereotyped. Another much-filmed writer, Leonardo Sciascia, commented that 'Sicily is cinema,' particularly when one considers the triangular island's frequent appearances as a movie backdrop.' (see Di Giorgi, Young, 1997 p. 20).
2. Gennari Santori, Paola (2003) *I finanziamenti dell'Unione Europea per l'Industria Cinematografica e Audiovisiva*. Pesaro: Mostra Internazionale del Nuovo Cinema, Antenna Media, Media Desk Italy. This booklet contains information on the procedures to access European funds.
3. De Grazia notes that today 'the contribution of cinema to national culture has grown in esteem, such that arguments on behalf of protecting European cinemas end up treating the cinema not as a mere commercial product but as a veritable monument to high culture, akin to opera, symphony orchestras or great masters' paintings, or alternatively as the artefacts of popular subculture indispensable to nurturing a multicultural identity within a "European homeland" that would extend from the Atlantic to the Urals' (De Grazia, 1998, p. 21).
4. 'The definition of what is intended by European identity and European culture remains blessedly open and inventive. There is reason to debate that there is such a thing as a European cinema, much less a European culture in any canonical sense. There is reason to denounce the dangers of invoking these legacies in the name of identity politics' (De Grazia, 1998, p. 29).
5. The (European) Media Commision sponsors Europa Cinemas, a theatre circuit composed mostly of art-houses or exhibition outlets that are expected to give priority to European films in their programming, and therefore receive different types of incentives to do so.

6. The website used for the test is http://rottentomatoes.com.
7. See Muscio, Giuliana (1999) 'Sceneggiatori e nuovo cinema italiano' in Marrone, Gaetana (ed.) *New Landscapes in Contemporary Italian Cinema. Annali di Italianistica*, The University of North Carolina at Chapel Hill, pp. 185–194.
8. 'Provincial cultures, however, were grounded in more local cultural forms, and were therefore, according to the definitions of the period, less 'sophisticated' (Higson & Maltby, 1999, p. 21).
9. The European Union certifies DOP products, that is, agricultural products that are typical of a specific geographic area, in a manner analogous to the denomination of wine origins.
10. Antenna Media of Turin surveyed this field for the European Coordination of Regional Investments Funds-Audiovisual, and reported that there are 148 funds supporting the audio-visual industry in Europe, and 77 have a regional character. These regional activities are coordinated by Wallimage and ECRIF-AV.
11. Sicilian cinema includes *1860* by Blasetti, *Stromboli* by Rossellini, *La terra trema,* and *Gattopardo* by Visconti, *Salvatore Giuliano* by Rosi, *L'avventura* by Antonioni, *In nome della legge*, *Sedotta e abbandonata* and *Divorzio all'italiana* by Germi, just to mention only some of the relevant titles. (See Genovese & Gesù, 1995.)
12. As Team Four of the ESF programme we have remarked that situations of extremely localized culture as the Basque region, could be associated instead with innovative – even experimental – forms of expression and modes of production, as the Guggenheim's experience in Bilbao confirms.
13. Crialese (Bertani, 2003) stated that he wanted to avoid Hollywood storytelling: 'I wanted something with a back taste of Italianity, but not folkloristic.' Thus the film was born out of a reaction to his American experience, but also avoiding the traps of cultural stereotyping: 'I wanted to be back in my old Italy' he says, 'not the urban one, which has changed a lot in the last ten years, but in that of my memory – the country I had been thinking of, while I was away.' Thus the film was born with the nostalgic look of the emigrant, who thinks about the 'old world' as a frozen experience, an image fixed in his own time. The a-historicity of the gaze transforms itself in 'mythological realism', as some Italian critics have defined it.
14. Interrogated on the application of *Respiro* for Eurimages, Procacci insisted that the main condition was the co-production, while he is not aware of any reading of the film script to assess its artistic qualities or its 'Europeanness'.
15. This wording immediately reminds one of the traditional discussion about national citizenship in Europe, and of its contradictions and differences. Citizenship is established according to either residence in a country or to birth (land), or according to the citizenship of the parents (blood). The slash dividing citizens/residents in the text points to this unresolved question.
16. http://rottentomatoes.com.

References

Beck, Ulrich (2003) *La società cosmopolita*. Bologna: Mulino.

Di Giorgi, Sergio, Young, Deborah (1997) 'The Dream of a New Sicilian Cinema' in *Cineaste* 13(1) pp. 20–23.

Bertani, Alessandro (2003) Emanuele Crialese (interview) in Cavandoli, Vincenzo (ed.) *Accadde domani. Nuovo cinema italiano*. Reggio Emilia: Ufficio Cinema.

De Grazia, Victoria (1989) 'Mass Culture and Sovereignity' in *The Journal of Modern History*. 61(1) pp. 53–87.

De Grazia, Victoria (1998) 'European Cinema and the Idea of Europe, 1925–1995', pp. 19–33 in Nowell-Smith, Geoffrey, Ricci, Steven (eds) *Hollywood and Europe. Economics, Culture, National Identity 1945–95*, London: BFI.

Gennari Santori, Paola (2003) *I finanziamenti dell'Unione Europea per l'Industria Cinematografica e Audiovisiva*. Pesaro: Mostra Internazionale del Nuovo Cinema, Antenna Media, Media Desk Italy.

Genovese, Nino, Gesù, Sebastiano (1995) *E venne il cinematografo: le origini del cinema in Sicilia*, Catania: Maimone.

Higson, Andrew & Richard Maltby (1999) 'Film Europe' and 'Film America'. *Cinema, Commerce and Cultural Exchange 1920–1939*, Exeter: University of Exeter Press.

Muscio, Giuliana (1999) 'Sceneggiatori e nuovo cinema italiano', pp. 185–194 in Marrone, Gaetana (ed.) *New Landscapes in Contemporary Italian Cinema. Annali di Italianistica*, The University of North Carolina at Chapel Hill.

Contributors' Notes

Karin Becker, Ph.D., Professor in the Department of Journalism, Media and Communication, Stockholm University and the Department of Culture Studies, Linköping University. She began her career in the mass communication and journalism programmes at Indiana University and the University of Iowa, specializing in documentary photography and photojournalism, and moved to Sweden in the mid-1980s. Her research focuses on cultural histories and contemporary contexts of visual media practices, in the press, in museums, in private settings and in ethnographic research. Her current research explores visual aspects of public space. Her English publications include *Dorothea Lange and the Documentary Tradition* (1980), *The Strip: An American Place* (1985), *Picturing Politics. Visual and textual formations of modernity in the Swedish press* (2000), and *Consuming Media* (co-authored, 2007).

Jérôme Bourdon (Ph.D., Institut d'études politiques, Paris, 1988) is head of the Department of Communications at Tel Aviv University and associate researcher with the Center for the Sociology of Innovation (CSI) in Paris. He is interested in the sociology of media professionals, the poetics of television genres, and the relations between social memory and television. He is completing a book on the coverage of the Israeli-Palestinian conflict in the western media, and working on the global history of television. Recent articles in English: "Live Television is Still Alive" (reprint in *The Television Studies Reader*, R. Allen and A. Hill, Editors, Routledge, 2004), "Public service television and the popular" (*European Journal of Cultural Studies*, 7/3, 2004). "Unhappy engineers of the European soul. The politics of pan-European broadcasting" (*Gazette*, 69/3, 2007).

Johan Fornäs, Ph.D., Professor at the Department of Culture Studies (Tema Q) and Director of the Advanced Cultural Studies Institute of Sweden (ACSIS) at Linköping University, Sweden; Vice Chair of the international Association for Cultural Studies. He has a background as Professor of Musicology at Göteborg University and then Media and

Communication Studies at Stockholm University, with research on popular music, youth culture, the Internet and media consumption. His English publications include *Cultural Theory and Late Modernity* (London: Sage 1995), *In Garageland: Rock, Youth and Modernity* (London: Routledge 1995), *Youth Culture and Late Modernity* (London: Sage 1995), *Digital Borderlands: Cultural Studies of Identity and Interactivity on the Internet* (New York: Peter Lang 2002) and *Consuming Media: Communication, Shopping and Everyday Life* (Oxford: Berg 2007).

Rob Kroes (Emeritus Professor of American Studies) is former chair of the American Studies programme at the University of Amsterdam. He received his Ph.D. in sociology from the University of Leiden (1971). He is a past President of the European Association for American Studies (EAAS, 1992–1996). He is the founding editor of two series published in Amsterdam: Amsterdam Monographs in American Studies and European Contributions to American Studies. He is the author, co-author or editor of 37 books. Among his recent publications are: *If You've Seen One, You've Seen The Mall: Europeans and American Mass Culture* (1996), *Predecessors: Intellectual Lineages in American Studies* (1998), *Them and Us: Questions of Citizenship in a Globalizing World* (2000), *Straddling Borders: The American Resonance in Transnational Identities* (2004), *Buffalo Bill in Bologna: The Americanization of the World, 1869–1922* (Co-author Robert W. Rydell, 2005), and *Photographic Memories: Private Pictures, Public Images, and American History* (2007).

Sabina Mihelj is a Lecturer in Media, Communication and Culture at Loughborough University. Her current research centres on issues of collective identity formation and mass communication, with particular reference to nationalism, European integration, and communism. Among her recent publications are "The European and the National in Communication Research," *European Journal of Communication* (2007), "'Faith in Nation Comes in Different Guises': Modernist Versions of Religious Nationalism", *Nations and Nationalism* (2007) and "Mapping European Ideoscapes: Examining Newspaper Debates on the EU Constitution in Seven European countries," *European Societies* (2008).

Giuliana Muscio is Professor of Cinema, head of the Master in Audiovisual and Multimedia Education and of the Master in Atlantic and Globalization Studies, at the University of Padua, Italy. She earned her Ph.D. in film studies at UCLA. She is author of *Hollywood's New Deal* (Temple University Press, 1996) and of works both in Italian and English on screenwriting, Cold War cinema, the 1930s media and politics, women screenwriters in American silent cinema, film and history, film relations between USA and Italy, Italian actors in Hollywood before WWII. She is the film consultant for the *Enciclopedia del cinema* published by the national encyclopedia Treccani.

Roger Odin (Emeritus Professor) was the Head of the Institute of Film and the Audiovisual Institute of Research at the University of la Sorbonne Nouvelle, Paris 3, from 1982 until June 2004. A specialist in communication studies, he is the author of numerous works with a semio-pragmatics approach to films and audio-visual productions, including two books: *Cinéma et production de sens* (A. Colin, Paris, 1990) and *De la fiction* (De Boeck, Bruxelles, 2000). He is also interested in documentaries (*L'âge d'or du cinéma documentaire: Europe années 50*, 2 volumes, L'Harmattan, Paris, 1997), home movies and amateur productions: *Le film de famille*, Méridiens-Klincksieck, 1995, *Le cinéma en amateur*, *Communications* no 68, Seuil, 1999. Recently (2007), he has organized an issue of *CiNéMAS* (Montréal) dedicated to the state of Film Theory: "La théorie du cinéma, enfin en crise".

Kevin Robins is Professor of Sociology, City University, London. He is the author (with Frank Webster) of *Times of the Technoculture* (Routledge, 1999) and co-editor of *The Virtual University?* (Oxford University Press, 2002). Recently he has been working on issues of migration in Europe, with a particular focus on Turkish migrants. He has also been involved in work relating to changing European culture and identity, and is author of *The Challenge of Transcultural Diversities* (Council of Europe, 2006). He is presently involved in the Goldsmiths Media Research Programme at Goldsmiths College, University of London.

Maria Rovisco is a Postdoctoral Research Fellow at the Department of Sociology, ISCTE – University of Lisbon. She gained her doctorate in Sociology from the University of York. In 2006 she was a Fulbright Visiting Fellow at the Center for Cultural Sociology at Yale University. She has published articles on the tradition of European films of voyage, symbolic boundaries and collective identity formation, and on cosmopolitanism and Europe's borders. She is currently working on a project on the arts and the public sphere in Portugal (1965–2005) and co-editing the book *Cosmopolitanism in Practice* (Ashgate). Her main research interests focus on film, cultural sociology, the arts and the public sphere, the idea of Europe, and social theory.

Philip Schlesinger is Professor in Cultural Policy and Academic Director of the Centre for Cultural Policy Research at the University of Glasgow. He has been a Jean Monnet Fellow at the European University Institute, a visiting professor at the University of Oslo, a visiting scholar at the Maison des Sciences de l'Homme Raspail in Paris, and has held the Queen Victoria Eugenia Chair at the Complutense University of Madrid. His research presently centres on literary ethnography, on creative industries policies, and on European communicative space. He has published widely on media and cultural questions. His most recent books are the co-authored *Open Scotland?* (Edinburgh: Polygon, 2001) and *Mediated Access* (Luton University Press, 2003). His most recent co-edited volumes are the *Sage Handbook of Media Studies* (Thousand Oaks, CA: Sage Publications, Inc., 2004), and *The European Union and the Public Sphere* (London and New York: Routledge, 2007). He is a Fellow of the Royal Society of Edinburgh and of the Royal Society of Arts and an Academician of the Academy of Social Sciences. He is a longstanding editor of *Media, Culture & Society*.

William Uricchio is Professor and Co-director of Comparative Media Studies at MIT in the US and Professor of Comparative Media History at Utrecht University in the Netherlands. He has held visiting professorships at Stockholm University, the Freie Universität Berlin, and Philips Universität Marburg, and has won Guggenheim, Fulbright, and Humboldt research awards. His broader research considers the transformation of media technologies into cultural practices, in particular, their role in (re-) constructing representation, knowledge and publics. Uricchio has written extensively on 'old' and new media, popular cultures, and their audiences. His current work takes up these issues through topics ranging from media historiography, to peer-to-peer communities, to computer games and history. His latest book is *Media Cultures* (Heidelberg, 2006).

LIBRARY, UNIVERSITY OF CHESTER